Controlling The Human Mind

The Technologies of Political Control or

Tools for Peak Performance

By Dr. Nick Begich

Earthpulse Press Incorporated
P. O. Box 201393
Anchorage, Alaska 99520 USA
www.earthpulse.com
Fax: 907-696-1277
Phone:1-907-249-9111

ISBN: 1-890693-54-5

Cover Art Design: Shelah J. Slade

First Edition, First Printing

Printed in the United States

Table of Contents

About the Author

Dr. Nick Begich serves as Executive Director of The Lay Institute on Technology, Inc. a Texas nonprofit corporation. He is also the publisher and co-owner of Earthpulse Press Incorporated, an Alaska-based organization.

Dr. Begich is the eldest son of the late United States Congressman from Alaska, Nick Begich Sr., and political activist Pegge Begich. He is well known in Alaska for his own political activities. He was twice elected President of both the Alaska Federation of Teachers and the Anchorage Council of Education. He has been pursuing independent research in the sciences and politics for most of his adult life. Begich received his doctorate in Traditional Medicine from The Open International University for Complementary Medicines in November 1994. Begich has also worked as both a Tribal Administrator and a Village Planner for the Chickaloon Village Traditional Tribal Council, a federally recognized American Indian Tribe of the Athabascan Indian Nation located in Alaska.

Begich co-authored with Jeane Manning the book *Angels Don't Play This HAARP; Advances in Tesla Technology.* He also coauthored *Earth Rising – The Revolution: Toward a Thousand Years of Peace* and *Earth Rising II – The Betrayal of Science, Society and the Soul,* both with the late James Roderick. Begich has published articles in science, politics and education and is a well known lecturer, having presented throughout the United States and in nineteen countries. He has been featured as a guest on thousands of radio broadcasts reporting on his research activities including new technologies, health and earth science related issues. He has also appeared on dozens of television documentaries and other programs throughout the world including BBC-TV, CBC-TV, History

Channel, Science Channel, TeleMundo and others. Begich has served as an expert witness and speaker before the European Parliament, GLOBE, the Alaska Legislature and many other organizations.

Dr. Begich is married to Shelah Begich-Slade and has five children. He resides just north of Anchorage in the community of Eagle River, Alaska, USA.

Dedication

This book is dedicated to Ms. Dorothy Lay of Dallas, Texas, founder of The Lay Institute on Technology Incorporated, a Texas based not-for-profit corporation. The development of these materials and their cataloging has been facilitated by the Institute in order to support the goal of educating the public on the impacts of technology in the 21st century and beyond. We have just completed the initial setup of the institute's internet website *www.layinstitute.org*. This site has a searchable bibliography with abstracts and significant public documents that speak to the issues described in this book, as well as many other areas where technology impacts humanity.

There have been many other people who, over the years, have significantly impacted my ability to continue in this work – most importantly my wife Shelah who has stood by me for twenty years. She has encouraged me, sacrificed beside me, rallied my spirits, held down the home front and been my best friend as we have engaged in our public outreach efforts on these topics.

There have been hundreds of people over the years who have contributed to this work by sending us excellent research material, financially helping when needed, encouraging us in countless ways and volunteering time on these issues and many others. I want to first thank those that are no longer among us in this life – James Roderick, Dr. Reijo Mäkelä and Dr. Gael Flanagan. These three all passed away over the last five years.

I also want to thank Harlan Girard, Duncan Roads, Art Bell, George Noory, Chuck Harder, Sabine Fechner, Robert Thiedemann, Katherine Wade, Chief Marie Smith-Jones, Chief

Gary Harrison, Nicholas J. Begich III, Patricia Wade, Michele Morton, Richard Alan Miller, Richard Lemieux, Tom Spencer, Stephen Rorke, Jeane Manning, Frank Polk, Charles and Lori Finkelstein, Glen & Betty Slade, Lucille and Dick Clark, the "Monkey Lads" and all of those who remain unnamed who helped along the way. We further dedicate this book to the victims of government abuse and, finally, dedicated to each reader who acts in the Spirit, towards a better humanity and better planet, as an individual act of stewardship.

It is through the combined effort of many that great things happen. We are all part of a great organic internet comprised of the human race in combination with all things under heaven. There are no closed systems. We are truly connected, and what each of us does has the potential to impact us all. So let's create carefully and make sure that the technologies we develop support the higher potentials of people, as individuals, created in the image of the Creator.

Part I

The Sinister Side of Controlling the Human Mind

Prologue

When I started my public work in 1993, I was still employed with the Anchorage School District. It was during that time that I began using articles, papers and other data I had accumulated over the twenty years as a springboard into the next decade of work. My work life at the time was no longer challenging. Most of my creative efforts, outside of the office, were focused on the things that were of significant interest to me – the science and public affairs.

I had gone to work for the District in 1984, in order to get a sense of local government and how it actually operated. I thought that would be a good place to start. It was the same place my father had begun his public work in the mid-1950's, before serving in the State Senate, and later the U.S. House of Representatives as a member of the 92nd Congress. The times were different then, but the conflicts were similar and much was gained in those years with the District. By 1993, I was considering where I was going in the institution and determined that I would stay and use my outside time to pursue my interests in science. I committed myself to connecting together interesting researchers in various fields of science while pursuing my own research in science and applied technologies. I felt that I was in a good position to do this, and it would be a valued contribution which would combine my interests in a more productive way. Never did I image what would follow.

The results of that decision have been mixed over the last twelve years in terms of ups and downs. Looking back I would not now choose a different path than the one followed because, even in the toughest times, a great deal was learned to equip me for the future. In each experience was an opportunity to learn.

In 1994 I wrote my first published article and the next year co-authored a book[1] on the HAARP project here in Alaska. With that effort we began a campaign of public disclosure on HAARP that continues to this day. There are more than 300 footnoted sources in that book that provide the foundation for the technology. The source material in our footnotes has been the hallmark of our work over the years in everything we publish. We have drawn from tens of thousands of documents in order to source the best material on each subject we cover. This present work is no exception.

By 1999 we had finished a good deal of work on HAARP disclosures and were committed to a number of projects that we felt would have a significant impact on the future of mankind. At the time, we were using our research to introduce new ideas and scientists to larger audiences. Our hope was that by combining technology reporting with radio discussion venues, lectures, and publishing that we could in fact raise awareness of otherwise obscure technical topics. We also felt that we could further the efforts of other researchers by helping them gain access to people who shared their interests through publication in popular venues.

Earthpulse Press, and those who have worked with us, has always seen its mission as translating science and technical ideas into plain language so that more people could learn and discuss the issues. Public education is what Earthpulse has always been about fundamentally. We have also utilized our communications to create support of other research efforts – all

1. *Angels Don't Play This HAARP: Advances in Tesla Technology*, by Jeane Manning and Dr. Nick Begich, 1995, Earthpulse Press Incorporated, Publisher.

of which have been sustained with the help of many people over the years.

In the course of disclosing the HAARP program we touched on the potential implications of technologies which could interfere with, or override, the mind and consciousness of people. The idea of manipulating the human mind for personal growth or for controlling others has been a theme for humanity for centuries. But now, with existing and quickly advancing sciences as well as the convergence of several technologies, controlling the mind and emotions is possible. The importance of the issue, in our opinion, is the most significant of our time. Should mankind interfere with the thoughts and emotions of people? Should anyone interfere with the free will of others? These are questions that this book asks. We each must answer, because it is this generation that will decide if these technologies enslave us or set us free to our higher potentials.

This book sprang out of a crossroads in life, a time when I was feeling that perhaps this was not the right work for me. I had lost three of my most important friends – James Roderick, Dr. Reijo Mäkelä, and Dr. Gael Flanagan. It was a tough time and by early 2004 I began to consider pursuing a much different path. Reijo and Gael had been great motivators behind a lot of my work and James had been my co-author on the Earth Rising series of books we researched together from the late 1990s through 2002. Jim had been working with me daily for four years until his death in August, 2002. My wife Shelah and I were tired of too often feeling that we were fighting the fight alone.

In the Spring of 2004 I was in the middle of storing a number of files and was actively seeking other possibilities and consulting projects that would have ended my work in these areas. It was then that I ran into a file, marked with Jim

Roderick's distinctive handwriting, a file named "Lay". I remembered the call we had received years before asking for a bit of information, which we always provided when we could. It was a request from one of the people who was working with Dorothy Lay several years before Jim had died. As I held the file I thought, "we need the help of someone like this who is interested in the areas of science that we report on if we are to stay in this work." At the same time, I made a small prayer that I would gain some level of clarity in the direction I should go in the coming days – to redouble my efforts in public education on technology issues or follow a new path. The one thing I knew for sure was we couldn't do it alone anymore.

Out of the blue, four days after finding the old file, the phone rang in my office. It was a friend of Dorothy Lay's calling to ask if I would consider being on Ms. Lay's Board of Directors for a nonprofit corporation she was setting up to help educate the public on technology-related issues. I was stunned. Was this the answer I was looking for? Of course I was interested, but at the same time cautious about any new relationships because of the experiences over the preceding years. I really felt that the timing was uncanny and that perhaps this was the answer that I was looking for in finalizing where my time would be spent in the coming years.

Dorothy and I spoke on the phone a few times in April and then again in late July, 2004, when we agreed to meet for a more in-depth exploration of the areas where we shared mutual interests. Even more importantly, we had the same motivation and interest in public education on technology as a foundation of cooperation. It was potentially an ideal combination if we could merge our efforts and impact the public discourse in a positive way. We met in August in Dallas for three days to discuss the possibilities and project interests we shared. We shared the same value system and interest in impacting the human condition. Over the next several months agreements were discussed and finalized which would initiate the work of the Lay Institute on Technology, Inc.

Throughout 2005 and 2006 I continued to lecture, research and compile materials as we put together the first projects sponsored by the Lay Institute and Dorothy Lay. The first project was the creation of the Lay Institute's website at www.layinstitute.org. The website will house the complete Earthpulse Press Index (EPI). We loaded it with seven-hundred document files in December, 2005. This index will serve those interested in these, and related topics, and it is our intention to continue to add to the database throughout the coming years. As a result of the Institute's efforts, we believe that we will be able to continue this work of public disclosure on technology for the foreseeable future.

Throughout the last twelve years we have continued to report and track trends in the area of controlling or influencing the human brain. By affecting the brain with existing and advancing technologies that we will explore in the coming pages, readers will see that these areas have perhaps the greatest implications of any on the future of humanity. The idea of controlling the mind – the thoughts, emotions, consciousness, and "idea centers" of people – is a subject that demands serious discussion, not just in secret government programs and science circles, but with the public. This book is intended to ignite an increased level of interest with the public in discussing these subjects. We will explore some of the technologies for enhancing human performance as well as those being developed that interfere with people in a manner previously only possible in science fiction.

Over the years Earthpulse Press had built a significant database of information on these subjects and in early 2004 produced a DVD *Mind Control* as well as a second DVD *Technologies of the 21st Century* where these issues are discussed. The combination of DVDs with written material, and an available database of information, provides a nexus where ideas can be opened to more popular discussion. So, it is hoped that this book, the new databases that we have helped develop with the Lay Institute on Technology, and presentations we make on radio, in lectures and in workshops will in fact serve the purposes of public education on a subject we consider one of the most important in the 21st century – *Controlling the Human Mind.*

Chapter One

The Root of the Technology

While writing this book, I was asked by one of my editors to provide a glossary to introduce the new vocabulary and ideas presented in these pages. I thought about how best to do this and felt that it would not be enough to provide just a glossary. The words needed to be connected in a way that really explained otherwise complex processes and concepts particularly, if they were being read for the first time, without a good deal of exposure to these specific areas of science.

The root of this technology is in thoughts, conveyed in words, which create thought-forms and images in the minds of readers. This is the essence of communication, which is in all specialized subjects built on the words used to provide the information. In the areas covered by this book, the words were created in the institutions of science and technology. The mastery of a few words will lead to new concepts and understandings that, once realized, will greatly facilitate the delivery of the information in the following chapters. Think about vocabulary built around things like lasers, radio, TV, telephones, home computers, or other technology that is now understood in our common knowledge. We all share the vocabulary of modern technology in these areas and will also in these new subjects.

One of my purposes, as an author, has been to find the best information and explain it in a way that is understood and communicated through the foggy language of technology and science. So, rather than a dry, disconnected glossary, it seemed a chapter to open this book should include a narrative explanation of the terms I will use. The key words are in bold for this chapter only in order to draw attention to the vocabulary that will be used as we review the science relevant to controlling the human mind.

The Energy of Life

Life consists of living **cells.** Inside each cell is **DNA** imprinted with the **genetic code** that controls every aspect of what we are as physical beings. The genetic code controls the development of the cell and production of **proteins** within the cells. The **proteins** provide structure to the cell and serve as part of the chemical processes that combine with food, producing the energy and the components needed for cells to continue to self-generate.

The body breaks down the foods we take into our bodies and captures the simplest molecules, and energy components, and then delivers them to the cells. This process of breaking down foods, and selecting the right molecules, represents the **chemical code** the body recognizes.

All of the chemical reactions in the cells are driven by electromagnetic oscillations, pulsations, vibrations or frequencies of the vibrating atoms and substances that make up their composition. This is the **frequency code** of all of atoms, molecules, cells, and components of any living organism. All physical matter is vibrating, energetically at some level, with the – *rate of vibration specific to the material just like the* ***genetic code*** *is specific to our physical make-up.*

The body only utilizes certain **organic molecules** in its processes. The unique **frequency codes** of those organic molecules serve as the mechanical switches and regulators for living organisms. Every place in the body that there is an exact **resonant frequency match** activated there is **resonance,** when their unique code is recognized the correct molecules are absorbed. All cells and cell groups have their own **resonant frequencies** built into their structure with the energy exchanges taking place on the surface of the membrane of each cell.

All physical matter is composed of systems in motion at their smallest atomic and subatomic levels. Everything in creation is composed of the same basic elements or building blocks, but in their unique combinations, they have the individual fingerprint of a **resonant frequency.** When a substance's natural vibration rate encounters other energy sources vibrating at the exact same rate, there is a transfer of energy between them that results in a biological reaction. The **resonating** material is coupled, or joined, with the energy source, which then directly impacts the targeted material. This will help explain how some of the new technologies described in this book operate when energy, in certain forms, is directed at people.

The **resonant frequency of a substance** is defined as its vibration rate under its normal and natural condition. When a material is activated by interaction with another source of energy at the same **resonant frequency** a more powerful and intensive response occurs. If energy is **pulse-modulated** into the substance a significant change can be created in the **codes** of the human body which is why **resonance** is such an import part of any of these discussions. **Resonance** is one of the significant keys to bridging the physics of materials science with organic chemistry, which will lead to the greatest breakthroughs in medical science in the 21st century.

Another very important discovery in both electronics and physiology were **liquid crystals.** In electronic circuits these types of crystals respond to energy so that they go from transparent to opaque or change their character in other ways. In living things **liquid crystals** also exist as **organic molecules** and have characteristics of both solids and liquids.

Inside and outside of each cell are **liquid crystal organic molecules** that will **resonate** with any outside source of energy where there is a **frequency match**. They will begin to vibrate at higher states of energy when an outside source is introduced, just like in electronics, when a radio station transmitter and home receiver **frequency's match**. Another example is when a singer hits a certain musical note. A substance, in this case a crystal glass, begins to vibrate in harmony, and sympathetically, with the **resonant frequency** of the singer's voice, at such a high enough energy level that it causes the glass to break into pieces. The energy charge on the inside and outside of cells can also be manipulated, in various forms, to change the **liquid crystals,** sort of like living micro-circuits.

All processes that build up and breakdown the cells and chemicals of the body are controlled by **electromagnetic oscillations.** The **metabolic processes** are processes that can be influenced through applied external energy sources of low power when they share the same **frequency codes.** When the laws of physics are applied to the materials that make up the human body a much different set of possibilities begins to emerge. Manipulation of the **frequency codes** of living things can change them in more direct and powerful ways than chemicals because the delivery systems are precise and only effect the targeted materials, elements, molecules, cells, organs, etc.

Electromagnetic fields can be introduced from devices outside of the human body with any specific living organism or individual substance in the human body being targeted. The

electromagnetic spectrum has within it ultra-low **frequencies** below a fraction of a pulsation per second up to significantly higher and higher **frequencies** including radio waves, light, microwaves, x-rays, cosmic rays, and so on. Any of these can be manipulated with significant effect to living creatures. The energy levels are significantly below the level of the earth's normal background radiation in many case, but the earth's are not **coherent signals**, rhythmic signals that can be understood by living things on an intensified level. These natural **signals** are random and generally not **coherent** or rhythmic in the same way as other biologically active signals. This is not to suggest that other, earth-driven cycles, or energy exchanges do not effect life as of course they do. What we are talking about here are specific energy sources of sufficient power and **signaling** capabilities to effect the **resonant frequency** of any substance.

What happens when an external electromagnetic field is in **resonance** with a biological molecule, then the same type of molecule will experience an energy exchange through **induced electron flow** and **electromagnetic coupling.** Researchers like Dr. Ross Adey, who will be introduced in coming pages, and others have shown that the cells of the body are like filters or tuners that only recognize a corresponding electromagnetic signal that **matches** with their own. **Electromagnetic coupling** allows the creation of very specific "controlled effects" over any aspect of a living creature. Once decoded, the understanding of the **frequency codes** of the body, brain and mind can be applied to people, in healing viral and bacterial intrusions, regulating chemical and metabolic processes in the body, inducing information transfers into the brain that alter perceptions and more.

When living things are also recognized as very complex living biophysical nanocircuits, and the laws of physics applied, very specific outcomes will always follow. There is a principle of **electromagnetic induction** where an electric current can be induced in a conductive material by just moving a magnet along the material. You can also measure the magnetic field created when a current flows through a conductive material. A **transformer** is a device that transfers electrical energy from one electric circuit to another through **magnetic induction** while the **frequency** stays constant. **Induction** is how energy is transferred in living things as well.

Another consideration has been the observation of weak **magnetic fields** in the human body and other living things. We have begun to hear words like **biomagnetic fields** in describing some observations about living things. We also hear discussion in health science, about **bioelectric** current flows through the body through our nervous system. As of only a few years ago, the bulk of the science community understood and accepted the "**chemical**" or "**biochemical**" model of living things and generally ignored the importance of the underlying physics.

The science of physics in the body, or **biophysics**, involves not only electrical currents in the body, but also the magnetic fields that are produced by those currents. An **electrical current** and a **magnetic field** go hand in hand, as one cannot exist without the other. More importantly, if one is impacted, the other is also affected in proportion. **Magnetic fields** were not detected in the human body until 1963 when researchers at Syracuse University measured the magnetic field of the heart at one-millionth of the strength of the earth's magnetic field.[2] In 1971 new equipment was developed that could measure the magnetic fields of the brain which were found to be 100 times weaker than the human heart's or 100 million times weaker than the earth's magnetic fields.[3] These

2. "Detection of the Magnetic Field of the Heart", GM Baule & R. McFree, Am Heart Jan, 1963, pages 95-96.
3. "Magnetoencephalography, Detection of the Brain's Electrical Activity with a Superconducting Magnetometer, Science 1972: 175:664-666.

and other new observations have added to the science. The field of **biophysics** is the more advanced science from which the most interesting breakthroughs are emerging based in creased understanding.

I had the opportunity to get to know Dr. Alexander Kaivarainen, the former Director of the U.S.S.R. Academy of Sciences, Biophysics Department, a few years after the collapse of the Soviet Union in the 1990's. These ideas were matters he was familiar with and confirmed with his research. These concepts, as they apply to living things, are not new and are being advanced in research circles around the world. Another friend, up until his death a few years ago, Dr. Reijo Mäkelä, a Finish research scientist and medical practitioner, demonstrated some of the fundamentals in **applied biophysics**. This is the idea of creating delivery systems which use the principles theoretically demonstrated by practical inventions. In applying the sometimes misunderstood principles of **biophysics** into the practical tools of electro-laser acupuncture, Dr. Mäkelä pioneered methods in health and stimulated thinking in adjacent fields as well. Since his death his work has continued to be advanced by his daughter Dr. Anu Mäkelä whose independent, and equally impressive, **applied biophysics** research is yielding incredible results. These individuals have succeeded in isolating many of the healing **frequency codes** of the human body and, importantly, are adding to a growing body of remarkably practical medical advancement toward the diagnosis and treatment of numerous disease states and conditions.

It is well known that high energy electromagnetic fields, or **ionizing radiation,** can cause heating, ionization and damage to living tissue. What is being learned, and better understood, is that the subtle energy of the body is much more interesting. It is at these lower levels of energy, or **non-ionizing radiation** levels, that the **resonance** effects of the

frequency code are found. It is in the subtle energy transfers between materials where we find the drivers of living things. Science is increasingly focused on these subtle sources of signaling energy flows through the body, as opposed to the traditional focus on the more powerful **ionizing radiation**, although this is still an ongoing research area.

Understanding the way that the body reacts to energy was also explored by others[4] and determined to be frequency-specific with respect to biological processes. By understanding the nature of a reaction that is frequency-specific we may tailor externally created signals that will be ignored and filtered out of the non-targeted biological material while having dramatic effects when the **codes are matched** to the targeted biological processes and components. This **coding** is the key to the health sciences moving away from partially failed ancient and current models of health. Moving from a **chemical model** of healthcare that is pharmaceutically based to one of energy based science will help avoid countless drug side-effects and the toxification of the body, which often results in permanent organ damage.

Frequency Modulation

Radio and television waves are created by the production of pulsing electromagnetic charges with each different station broadcasting on a specific **frequency**. This **frequency** is always the same and is where the radio or television station is on the dial. It is called the **carrier wave**. It is the information transferred on a **carrier wave** that is translated by the electronics of a radio or television receiver as either sound or sound and picture. Over one hundred years ago Nikola Tesla discovered that a **carrier wave** could be used to carry other

4. "Critique of the Literature on Bioeffects of Radio frequency Radiation: A comprehensive Review Pertinent to Air Force Operations, Final Report USAFSAM-TR-87-3, June, 1987.

signals called **signal waves**. A **signal wave** is what actually delivers the electronic **code** that the electronic circuits sort out and deliver as the image or sound.

When a **signal wave** is placed on a **carrier wave** we get what is called **modulation. Modulation** can be thought of as a small **pulsation** that either effects the height of the wave, also known as **amplitude,** or effects the distance between pulses, represented as pulses per second (**hertz**), or what we call **frequency**. What makes radio and television work is the combination of the **carrier wave** with a **signal wave**. When these are received by a radio or television set the combined signal is sent through the electronics of the receiver and projected on a screen as an image, or through a speaker as sound. The human body also translates external signals through its **biocircuits** in the same way.

We can also think of the whole body, an organ, cell, molecule, element or atom as a **transducer** (converting energy) and **a receiving antenna** (receiving energy) tuned to the exact **signal wave** on a **carrier wave**. When a **receiving antenna** picks up a signal of a broadcast station, and a circuit is tuned to that signal, **resonance** occurs and the received signal is increased in signal strength through **amplification** by the electronic circuit. Again, when we look at the human organism, which is constructed of what are essentially cellular **biological oscillators**, we see that the body can translate the information in the **signal wave** and transfer energy from the **carrier wave** through the same laws of physics applied to radio and television. These understandings are leading to "controlled effects" where applied electromagnetic fields may influence or disrupt biological processes, emotions, and senses.

The "controlled effects," sought by the United States Air Force and others as revealed in their own publicly released documents and described in detail in the following chapters, attempts to influence all of the senses, including sight, sound, taste, smell, touch and those additional senses and capabilities that are anomalous or rare in people. In addition, research will be revealed that shows the same progress in effecting the mind and any substance or collection of materials in the body. Over fifty years, a whole revolution of weaponry has been advanced, which is based on these new ways of impacting or degrading human health.

Researchers are now showing that **frequency modulation** of cell membrane receptors, which function as **antennas/transducers,** transfer signals that are understood by the cells. Researchers have shown that all physiological processes from metabolic functions, nerve impulses and even thoughts are defined by their internal **codes,** which dictate how they interact with other energy sources of many kinds. The greatest advances are represented in the increased understanding of the **genetic codes** relationship to the **chemical codes,** and the **biophysics** of the **frequency codes** of living systems. When these areas of science merge and their laws are contiguously applied to our observations everything will change: health, the mind, and even consciousness itself.

The Body as a Subatomic-Nano-Micro Circuit

In the past, it was assumed that chemical reactions, in the fabric of molecules, direct the flow of **signals** and the **transfer of energy** in living systems. Instead this old idea has been shown to be driven by underlying processes of **energy exchange** at the **atomic** level, and below, that then affect the chemistry of all living things.

Within living cells proteins exist that have specific three-dimensional physical characteristics and forms. The form of **proteins** arise out of their unique order of **amino acids**,

electrical charge and **polarity**. Sequences of **amino acids** have been found to **coil** or **wind themselves into a helical spiraling shape** called an alpha-helix. In electrical terms **coils** and helices are **inductors, transducers and antennas**. These can change shape and act as on-off switches or perform **binary functions,** which are required for any computer to work whether it is your desktop or is the human body functioning as a supercomputer. These **coils and helixes** are **biological circuits** serving the same kind of functions as inorganic, nonliving electrical circuits.

It has been shown that by adding energy directly into the human body, in the right forms, the overall energy level of the body can be effected both positively or negatively. What is being discovered in these biophysical processes is what we are now referring to as "nanotronic nutrient" effects. A "nanotronic nutrient" is either a device or a dietary compound that serves as a delivery system for **frequency modulated effects**. An example would be the use of five nanometer gold particles combined with insulin and aspartic acid, an amino acid. "The acid produces a charge that allows the insulin to adhere through electrostatic interactions." The result was a nasal spray that reduced blood sugar levels 55% in two hours in tests on rats. "The method could emerge as a new platform for delivering drugs that otherwise must be injected because they break down in the stomach."[5]

There are many ways to get the right things to happen. In living things it is possible to induce energy transfers which result in improved health because of the way that organic molecules as nutrients, trace elements and other substances are utilized by living systems. *The use of light, sound, microwave, radio, television networks, computer systems or power grids, all of through which a* ***pulse modulated signal*** *can be*

5. "Nanotech, New Way to Sniff Drugs", MIT's Technology Review, May/June, 2006, pg.30, EPI6072

introduced, can be used to create a deliberate effect, or a side effect, impacting living organisms in neutral, positive or negative ways. Because research has shown that energy exchanges, of very specific types, drive the foundational processes of life itself there is a need to understand how these signals work and design systems that avoid any accidental or purposeful biological effect without the clear consent of people. There are real possibilities that are already advancing without any regulation or control to protect people from the wrongly applied effects of these technologies. At the same time, there are real possibilities that through convergence of several technologies great breakthroughs will be made in longevity, health, and increased human potential. Unless we master our understandings in these areas others will continue to expose people, the environment and every living thing on the planet to forces that can be damaging and to some deadly. There is increasing evidence that not only the chemical soup of mankind but, also the electromagnetic mix we have created, may be contributing more to stress and other health disorders than once imagined, destroying the environment and creating problems that will persist for decades to come.

It is the pulsing flow of energy that regulates the human body and all living and nonliving physical things. Mankind will soon be capable of transforming life by creating the right subtle flows of energy that drive all living things. There are incredible possibilities and opportunities as these technologies advance and we develop increasingly targeted and specific uses.

This book is divided into two parts. The first part reviews the history of the technology we explore, which unfortunately has been predominantly driven by military research funds supporting espionage and warfare uses. This will introduce readers to some of the most advanced projects, while signaling a call for all of us to carefully consider the implications of our science and technology. Moreover, we expect that this will invigorate the debate regarding these areas of science. The second part of this book begins to explore the possibilities of the technology when applied toward enhancing human potentials. This is where we may begin to recognize our vastly greater possibilities as individuals through tools and techniques that are available today. This is the century of the brain, which ultimately is about the mind and consciousness itself. We are opening the gateway to a deeper understanding of the essence of who we are, who we will choose to be and who future generations will become.

Chapter Two

Controlling the Human Mind

On the Way to *1984*

I have been writing about technology areas since 1994 when *Nexus Magazine* published my first article on HAARP – a huge array of radio frequency broadcasting antennas operated by the Navy and Air Force here in Alaska. There were aspects of that project that were very disconcerting including the creation of extremely low frequency (ELF) pulses of energy that are known to interact with humans and animals. Moreover, the device could be manipulated in such a manner as to impact every vital organ of the human body if operated by a skilled person with the intent to do so, or as a "side effect" of other military uses for the HAARP technology. In the book, co-authored with Jeane Manning, *Angels Don't Play This HAARP,* we discussed in depth the HAARP project with all of its implications, including the issues of emotion and mind control.

The idea of controlling the human mind goes back before the HAARP research. There was a lot happening in the 1970's and 1980's in the field of brain research. The ideas of enhancing learning, curing brain dysfunction, affecting emotions and, by controlling the human brain, perhaps impacting the entire body, were the subjects of numerous press reports and academic studies. I was researching these areas throughout the 1980s in conjunction with my varied interests at that time. The idea of influencing human events by direct control of the processes that control the bodies and brains of people is the subject of this book which resulted.

Controlling the Human Mind grew out of two very strong interest areas on the part of the author – enhancing human performance on one hand and the potential misuses of these technologies on the other. The ability to enhance human performance and increase human capacity is quickly becoming a necessity in an increasingly complex world. People are gaining in knowledge and new science seemingly without end. Today technology is advancing so fast that what was previously impossible is now possible. Technology, from the invention of the wheel to where we are today, is doubling every nine to ten months.

What has been disclosed by the military are weapons which interact with our living energetic systems – living systems which are responsible for sustaining our lives and mental processes. These new weapons are unlike anything ever contemplated by mankind. These are weapon systems which pierce the very integrity of the human being. In these pages over 280 source documents trace the evolution and use of some of these advances. I am sure that what I found in the open literature represents a very small part of what has been developed. I am hopeful that this text will be revised from time-to-time to include additional details as the information is collected through the Lay Institute's efforts.

On the surface, it is easy to see why some people would desire to subdue rather than kill using the new systems being developed and deployed by the military today. On the surface it seems that some "minor" violation of civil rights might be overlooked in favor of maintaining existing social orders. On the surface it might even seem desirable to some people that controlling the behavior of others, in order to achieve political ends, is acceptable. The ethical line has not been drawn in terms of the use of these new technologies. The reason this line has not been clearly drawn is because science is advancing at ever-accelerating rates while the proper ethical discussion is lost in the march by militaries in their new revolution. I am suggesting looking back toward those traditional values from which all of the world's cultures and societies have emerged. I propose a revolution which will awaken each of us to the reality of our shared values and hopes for future generations through education and responsible action. I suggest a revolution which is focused in peace, which holds the individual human as the central actor in achieving that peace and stands firmly against violations of the rule of law and individual human dignity.

The issues uncovered in the research of new technology are frightening in their implications. We have uncovered over forty United States patents dealing with the control and manipulation of the mental and emotional states of humans. We have culled from thousands of government, academic and major media reports the evidence of these new technologies in terms of their potential effects on us. We trace the roots of this research back to the early 1950's, when major initiatives began to be made by the military to control human behavior.

The early attempts used chemicals and hallucinogenics to achieve some measure of control. Then in the early 1960's the interest changed to non-chemical means for affecting behavior. By the early 1970's, within certain military and academic circles, it became clear that human behavior could be modified by the use of subtle energetic manipulations. By 2006, the state of the technology had been perfected to the point where emotions, thoughts, memory and thinking could be manipulated by external means.

Stop for a moment and consider the impact of what this means – the idea that human thinking can be disrupted or manipulated in a way that can't be resisted. The ability to impact thinking in this way comes through, according to a leading Russian scientist, Dr. Igor Smirnov, "as if it were a commandment from God, it can not be resisted."[6] Think about this for a moment. What else could be a greater violation of our own individual personhood? These new systems do not pierce the tissue – they violate the very essence of who we are – they violate the internal and private aspects of who we are as individuals. The idea that some external force can now disrupt not only our emotional states, but our health as well, should not come as such a great surprise. What is surprising is the volume of evidence now available to us that documents that these technologies are here now.

One of the most revealing sources we found regarding these new technologies was produced by the Scientific Advisory Board of the Air Force. The Air Force initiated a significant study to look forward into this century and see what was possible for new weapons. One of those forecasts shockingly revealed the following:

> "One can envision the development of electromagnetic energy sources, the output of which can be pulsed, shaped, and focused, that can couple with the human body in a fashion that will allow one to prevent voluntary muscular movements, control emotions (and thus actions), produce sleep, transmit suggestions, interfere with both short-term and long-term memory, produce an experience set, and delete an experience set."[7]

6. *Undercurrents*, CBC-TV, Canada, February, 1999.
7. USAF Scientific Advisory Board. *New World Vistas: Air and Space Power For The 21st Century - Ancillary Volume.* 1996, pp. 89-90. EPI402

Contemplate this for a moment – a system that can manipulate emotions, control behavior, put you to sleep, create false memories and wipe old memories clean. Realizing this was supposed to be a forecast should not cause one to believe that it is not a current issue. These systems are far from speculative. In fact, a great deal of work has already been done in this area with many systems already in existence. The paper went on to say:

> "It would also appear possible to create high fidelity speech in the human body, raising the possibility of covert suggestion and psychological direction. When a high power microwave pulse in the gigahertz range strikes the human body, a very small temperature perturbation occurs. This is associated with a sudden expansion of the slightly heated tissue. This expansion is fast enough to produce an acoustic wave. If a pulse stream is used, it should be possible to create an internal acoustic field in the 5-15 kilohertz range, which is audible. Thus, it may be possible to 'talk' to selected adversaries in a fashion that would be most disturbing to them."[8]

Is it possible to talk to a person remotely by projecting a voice into his head? The authors suggest that this would be "disturbing" to the victim, what an understatement – it would be pure terror. A weapon which could intrude into the brain of an individual represents a gross invasion of one's private life. The idea that these new systems will be perfected in the next several years should be cause for significant discussion and public debate.

8.USAF Scientific Advisory Board. *New World Vistas: Air and Space Power For The 21st Century - Ancillary Volume.* 1996, pp. 89-90. EPI402

The Mind Has No Firewall[9]

In another article published in the Spring, 1998, edition of *Parameters*, U.S. Army War College Quarterly, an article by Timothy L. Thomas appeared – "*The Mind Has No Firewall.*" The article was perhaps the most revealing in terms of what can be expected in the future.

For decades, the United States, the former Soviet Union, and others have been involved in developing sophisticated new systems for influencing human physical and mental health. The desire and focus of this research has been to discover ways of manipulating the behavior of humans to meet political ends in the context of war-making and defense. What is interesting in all of this is the sophistication of external devices which can alter our very nature. In the article "The Mind Has No Firewall" the author states:

> "A recent Russian military article offered a slightly different slant to the problem, declaring that 'humanity stands on the brink of a psychotronic war' with mind and body as the focus. That article discussed Russian and international attempts to control the psycho-physical condition of man and his decision-making processes by the use of VHF-generators, 'noiseless cassettes,' and other technologies."

The article goes on to describe that the aim of these new weapons is to control or alter the psyche or interfere with the various parts of the body in such a way as to confuse or destroy the inner-body signals that keep the living system operational. The article describes the way "Information Warfare Theory" neglects the most important factor in information warfare – the human being. Militaries publicly

9. "The Mind Has No Firewall", U.S.Army War College, *Parameters,* Spring 1998, pp.84-92. EPI6032

focus on hardware and software, neglecting the human "data-processor". In the information warfare theories put forth in the past, discussion was limited to manmade systems and not the human operator. Humans were considered in information warfare scenarios only in that they could be impacted by propaganda, deceit and deception – tools recognized as part of the military mindset and arsenal. However, there is a more sinister approach, an approach which must be considered in the context of basic human rights and values which are fundamentally and foundationally based on our right to think freely. The article went on:

> "Yet the body is capable not only of being deceived, manipulated, or misinformed but also shut down or destroyed – just as any other data-processing system. The data the body receives from external sources – such as electromagnetic, vortex, or acoustic energy waves – or creates through its own electrical or chemical stimuli can be manipulated or changed just as the data (information) in any hardware system can be altered."

The aim of any information war ultimately deals with human beings. The policy of the United States is to target all information dependent systems "whether human or automated", and the definition extends the use of these new technologies to people – as if they were just data-processing hardware.

The *Parameters* article went on to discuss the work of Dr. Victor Solntsev of the Baumann Technical Institute in Moscow. He insists that the human body must be viewed as an open system instead of simply as an organism or closed system. This "open system" approach has been held by many Russian researchers and others going back to at least the early 1970's, according to documents held by Earthpulse. What is interesting is that it has taken thirty years to be seen in the open literature as a credible view of reality. Dr. Solntsev goes on to suggest that a person's physical environment can cause changes within the body and mind whether stimulated by electromagnetic, gravitational, acoustic, or other stimuli.

The same Russian researcher examined the issue of "information noise" which can create a dense shield between a person and external reality. The "noise" could be created as signals, messages, images or other information with the target being the consciousness of the group or individuals. The purpose would be to overload a person so that he no longer reacts to external stimulus or information. The overloading would serve to destabilize judgment or modify behavior.

According to Solntsev, at least one computer virus has been created which will affect a person's psyche – "Russian Virus 666". This virus appears in every 25th frame of a computer's visual display where a mix of color, pulse and patterns are reported to put computer operators into a trance. The subconscious perception of the display can be used to induce a heart attack or to subtly manage or change a computer operator's perceptions. This same system could be used in any television or visual broadcast.

In a July 7, 1997, *U. S. News and World Report* article, it was revealed that scientists were seeking specific energy patterns which could be externally applied to the body of individuals for the purpose of modifying their behavior. The article addressed some of the important public revelations about these new systems. These "revelations" represent but a small part of the story.

Since we first discovered these materials we have pieced together the bits and pieces of information that weave a more complete story. What we now know is that the most powerful weapon systems will be those which subdue populations through the use of subtle energy fields and trick the body and mind into reacting as if the signals were normal and natural. These systems can also be used to create total disorientation in people and cause "mystery" illnesses as well as be used to induce, at a distance, heart failure or significant respiratory distress, among other things.

One of the other important areas of my current work deals with the issues of privacy. Under the guise of such thinking as, "if you don't have anything to hide, why are you concerned? This is for your own safety and protection" – comes a theme voiced by every repressive regime ever to emerge out of human interactions. Fear is being used to dismantle our system of laws and civility. International drug traffickers and terrorists are the first targets of these new technologies, based on the idea that the average person will look the other way when the rights of these suspects are considered. However, except in the most extreme situations, the first questions which should be raised deal with the idea of innocence until guilt is proven, concepts of due process of law, the right to a reasonable defense, and a fair trial. Who is the guiding hand that will otherwise decide the guilt or innocence of suspected persons? Should we look the other way or should we insist that the values on which all democratic constitutional law is based are equally enforced? Recent developments on the privacy front are summarized below:

• The United States government, in cooperation with several other governments, now monitors all forms of electronic communications including telephone, fax transmissions and internet communications. This is done through huge computer systems that pull key phrases or words out of all of these communications and then deliver the information to numerous security agencies in the participating countries. To get around each country's domestic privacy laws another country does the spying and transfers the data to interested parties. Part of this system is called Echelon.

• It is now possible to follow the movements of any individual anywhere through their cell phone. In fact, in the United States, under the guise of locating people in emergency situations it is now required that all cell phones have this tracking ability. It is also possible, with this new technology, to use microcircuits in these phones just like a hidden microphone to monitor the conversations of cell phone owners even when the device is switched off. The same systems are installed in new automobiles as well.

There is also a new directional cell phone tower that was announced by Nokia Research Center in Helsinki, Finland. This new design will increase network capacity three fold by changing from a shared signal from a cell tower that is spread over up to three kilometers, to a focused beam of one square meter.[10] It also reduces the power demand by using less energy to hit the target. Think about this for a moment. The cell phone system could be changed to one that can target any individual with a carrier signal that will only hit one individual.

• In many urban areas micro-cameras are being installed inclusive of microphone pickups in order to monitor "criminal activity". These cameras are being installed in numerous major urban areas and, based on miniaturization and decreasing technology costs, it will soon be possible to literally monitor entire countries. Cameras are so sophisticated that they can automatically track and read a vehicle license plate at 60 miles an hour.

• Sophisticated computer systems are under development which are designed to read body language in order to interpret potential violent or other behaviors and dispatch authorities to the scene.

10 . "Telecom, A cell-Phone Tower with Focus", By Wade Roush, *MIT's Technology Review*, May/June, 2006, pg. 22, EPI6073

• The ability to look into the homes of individual people, using very high-tech infrared cameras and other technology combinations, is now available to law enforcement. These technologies allow people to literally look into the bedrooms and private activity of anyone.

• The ability of televisions, through cable connections, to be both transmitter and receiver is not far away and is being tested and introduced as new burglar alarms and home safety devices.

• The technology to look through people's clothing to find hidden weapons or drugs is now in use by some policing and military organizations.

• Roving wiretaps which involve the idea that whole areas of cities can be tapped in order to find a single person who might be engaged in criminal activity. It is the same logic behind collecting DNA samples to prove an individual's innocence or getting fingerprinted at the bank in order to cash a check – all are gradual intrusions into privacy.

• Genetic predisposition-to-violence studies which involve the concept that certain genetic markers can be identified showing who is likely to exhibit criminal behavior, are now ongoing. Once identified, the individual can be medicated with mood altering drugs in advance of such anticipated behavior. During Bush Sr.'s administration, this program was almost implemented in urban school systems in the United States. The $400,000,000 United States initiative was stopped when those engaged in real science began to object. Now, over a decade later, this issue is again surfacing, the interesting point being that the same indicators in terms of personality for criminals are the same for non-conforming elements of society that produce the leadership within countries.

• Mandatory bank transaction reports including "reporting suspicious activity" are also being initiated in a number of countries around the world.

These are but a few of the things that are happening now and are presented here only to give indications of the general direction of these invasive technologies.

Psycho-Terrorism

The term 'psycho-terrorism' was created by Russian writer N. Anisimov of the Moscow Anti-Psychotronic Center. He indicates that psychotronic weapons can be used to take away part of the information which is stored in a person's brain and send it to a computer which reworks it to the level needed to control the person. The modified information is then reinserted into the person's brain and thought by them to be their own information. These systems are then able to induce hallucinations, sickness, mutations in human cells, zombification or even death. These technologies include VHF generators, X-rays, ultrasound and radio waves. Russian army Major I. Chernishev described in the military journal *Orienteer* (February,1997), how "psy" weapons are under development all over the globe. Specific types of weapons he noted were:

• A psychotronic generator which produces a powerful electromagnetic emanation capable of being sent through telephone lines, TV, radio networks, supply pipes and incandescent lamps. This signal would manipulate behavior of those in contact with the signal.

• A signal generator that operates in the 10-150 Hertz band which when operating in the 10-20 Hertz range creates an infrasonic oscillation that is destructive to all living organisms.

• A nervous system generator that is designed to paralyze the central nervous systems of insects. This system is being refined to have the same effect on humans.

• Ultrasonic signals of very specific design have been created. These devices are supposedly capable of carrying out bloodless internal operations without leaving a mark on the skin. They can also be used to kill.

• Noiseless cassettes have been developed by the Japanese which has given them the ability to place infra-low frequency voice patterns over music, patterns that are detected by the subconscious. The Russians claim to be using similar "bombardments" with computer programming to treat alcoholism and smoking.

• The 25th-frame effect discussed above is a technique where every 25th frame of a movie reel or video footage contains a message that is picked up by the subconscious so as to alter the conscious mind.

• Psychotropics are defined as medical preparations used to induce a trance, euphoria, or depression. These are referred to as "slow-acting mines". Symptoms could include headaches, noises, voices or commands in the brain, dizziness, pain in the abdominal cavities, cardiac arrhythmia, or even the destruction of the cardiovascular system.

What is written here is the tip of a very large iceberg. These bits of information are intended to draw your attention to the state of the technology and where it is going. The conclusions are not based on speculation but, rather, are based on the facts presented by military and academic researchers from the United States and around the world.

One of the roots of the problem deals with free flowing debate of weapons concepts. The greatest issue moving against humanity is the "secrecy syndrome" which governments have succumbed to since the end of World War II. This government paranoia is the enemy of freedom. We need to recognize that within democratic institutions the requirement to disclose weapons and technology concepts should be a matter for public debate. We do not have to tell people how to build these systems, but we must be able to debate these new concepts.

Strange Science of the CIA

The military's interest in mind control goes back to Korea in the 1950's. Right after the Korean War, returning veterans and prisoners of war exhibited very unusual behavioral changes in terms of their beliefs and personalities. As our military began to look at these men, they discovered that huge psychological transformations had been made through methods used by the North Koreans and their Chinese allies. In 1956 the following was written into the United States Congressional Record:

> "Reports of the treatment of American prisoners of war in Korea have given rise to several popular misconceptions, of which the most widely publicized is 'brainwashing.' The term itself has caught the public imagination and is used, very loosely, to describe any act committed against an individual by the Communists. Actual 'brainwashing' is a prolonged psychological process, designed to erase an individual's past beliefs and concepts and to substitute new ones. It is a highly coercive practice which is irreconcilable with universally accepted medical ethics. In the process of 'brainwashing,' the efforts of many are directed against an individual. To be successful, it requires, among other things, that the individual be completely isolated from normal associations and environment."[11]

11. U. S. Senate. *Communist Interrogation, Indoctrination and Exploitation Of American Military, and Civilian Prisoners.* Committee on Government Operations, Subcommittee on Investigations. 84th Congress, 2nd Session. Dec. 31, 1956. EPI1131

Isolation is used by the United States in places like Guatánamo Bay or the other offshore locations is well known[12] . The U. S. uses these locations to conduct interrogations that last for over three and a half years without charges, legal council, hearings, trials or the application of civil law. On June 29, 2006 the United States Supreme Court struck down the Presidential Order that allowed the misuse of Presidential power to go on for almost four years.[13] The Court ruled that the Bush Administration had exceeded its authority under U. S. law and was in violation of the terms of the Geneva Convention.[14] In other locations torture, humiliation and other inhuman practices have been increasingly reported in the "War on Terrorism".

As a result of what was going on in North Korea in the 1950's, our military and others became very interested in the idea of manipulating human behavior for many different purposes. Obviously in prisoners-of-war situations the idea of interrogating people so they are more likely, and susceptible, to give up information was the main emphasis during the Korean conflict and, of course, was of interest to our military since.

From the 1950's things changed dramatically and the United States began to study the area of mind control through a number of classified programs. As we moved into the 1960's a lot of this interest migrated toward the idea of electronically controlling what happens within the human brain.

As far back as 1952[15] the United States government had actively engaged in exploiting extrasensory perception (ESP)

12. "The Experiment", by Jane Mayer, *New yorker,* July 11, 2005, Vol. 81, Issue 20, pg 60 (11 pges), EPI6067
13. "Wartime Policies Thrown in Doubt", by Jonathan Landay, Marisa Taylor and Margaret Talev, Mc Clatchy Newspapers, Anchorage Daily News, July, 1, 2006, pg 1, EPI6074
13."Court Sets Detainee Trial Rules", by Charles Lane, *The Washington Post,* June 30, 2006, reprinted Anchorage Daily News,EPI6064
15. Office Memorandum, United States Government, January 31, 1952. EPI6001

for military uses. Through the idea of using mental telepathy, out-of-body travel (remote viewing) and stimulating other anomalous human capabilities it was believed that new intelligence and military capability could be gained. There was a race to find out if these human attributes or anomalous capabilities could be exploited for military and national advantage. The Central Intelligence Agency (CIA) was the main organization pursuing these areas in the United States government. While these areas were closely related to the issues of mind control, more recent literature refers to these unusual abilities people sometime exhibit as "anomalous human capacities".

A look at these old documents sounds a familiar chord in 2005 and 2006 as we begin to hear new disclosures of CIA abuses and even torture[16] as a means for gathering intelligence in the case of international terrorism. One of the other interesting points made in the 1950's records is the reference to "overseas operations" outside of the legal restrictions of the United States – too often a reoccurring theme on the part of the shadowy aspects of the U. S. government in their effort to gain intelligence.

One of the earliest U.S. projects was Blue Bird, later changed to Artichoke, which was run by the CIA with an emphasis on the use of hypnosis and narcotics[17]. The project included extensive use of hypnosis with the intent to find ways to enhance interrogations, create amnesia and develop a "manchurian candidate". LSD and other drugs were administered with these outcomes in mind. Even the use of lobotomy as a means for controlling human intelligence was considered[18]. Through 1955-1956 knowledge was further developed for using various drugs in influencing hypnotizability

16. CIA Free to Export Prisoners, Anchorage Daily News, Page 1, March 6, 2005. Reprint from the New York Times by Douglas Jehl and David Johnson. EPI6053
17. Memorandum for The Director of Central Intelligence, July 14, 1952. EPI6002
18. Office Memorandum, United States Government, February 12, 1952. EPI6003

and immunity from it, and research was slated to continue until at least 1967,[19] through a succession of projects, including the infamous MKULTRA affair run by the CIA. One of the participants in these experiments was none other than Theodore Kaczynski a.k.a. "The Unibomber".[20] Kaczynski was a volunteer in the mind control experiments sponsored by the CIA at Harvard in the late 1950's and early 1960's. In the article about him, the use of LSD and other materials was described, raising the question of whether or not this person's problems may have started with his stint as a volunteer.

The article[21] goes on to discuss concerns with the modern drugging of America through the use of Prozac and Ritalin which are now, by some researchers, being associated with psychotic episodes in children and adults. The problems associated with the drugging of American youth was a long time in coming and started with as much controversy in the beginning as there exists today. If one looks back to the Congressional Record of September 19, 1970[22] much is revealed. The hearing was on the increased use of drugs in treating certain mental and other conditions in children. This hearing was in advance of the massive drugging of American youth taking place through the last three decades. This is a huge program for controlling behavior through the use of drugs rather than other means and is now a multibillion dollar enterprise.

There were attempts made under the Freedom of Information Act to force the CIA to disclose the names of the

19. Studies of Dissociated States, released under FOIR, undated. EPI6004
20. *We're Reaping the Tragic Legacy From Drugs*, by Alexander Cockburn, Los Angeles Times, July 6, 1999. EPI6020
21. Ibid.
22. Federal Involvement in the use of Behavior Modification Drugs on Grammar School Children of the Right to Privacy Inquiry, Hearing Before a Subcommittee of the Committee on Government Operations, House of Representatives, Ninety-first Congress, Second Session, September 29, 1970, US Government Printing Office, 1970. EPI6021

researchers and the institutions that were involved in the MKULTRA projects. This was even taken to the United States Supreme Court, which affirmed the right of the CIA to withhold this information under National Security.[23]

A "Manchurian Candidate"

The idea of a "manchurian candidate" being created can be traced back to the research of Dr. George Estabrooks who worked in Morton Prince's lab at Harvard University in the 1920's. He believed that one could use hypnosis to create a multiple personality for use in developing a super spy.[24] He pushed this idea on the U. S. military throughout the 1920's without success. However, after an incident in the Soviet Union the research in this area became of interest to the military. Estabrooks' work was classified by the mid-1930's. Apparently in his archives it was noted that he stopped publishing on these subjects at that time.

A review of the work of his mentor, Morton Prince, is interesting as it appears in the index of the Library of Congress. The 569 page book, *Dissociation of a Personality* by Morton Prince, published in 1906, remains a classic in psychology. Prince also wrote an earlier book in 1885, called *The Nature of Mind and Human Automatism.* Estabrooks' interest was focused on the idea that hypnosis could be used for creating the perfect spy[25] and by 1963 he believed that he could create a multiple personality and use them as he saw fit. He said, "this is not science fiction, it is fact. I have done it."

23. United States Supreme Court. 471 US 159, 85L Ed 2d 173,105 S Ct 1881, Case Numbers 83-1075 and No. 83-1249, Argued December 4, 1984. Decided April 16, 1985. EPI6071
24. http://www.mindcontrolforums.com/radio/ckln19.htm (This site has a number of materials that would be of interest to researchers on these subjects.) Lecture by Dr. Alan Scheflin, Professor of Law at the University of Santa Clara, The History of Mind Control: What We Can Prove and What We Can't. EPI6029
25. Hypnosis Comes of Age By G.H.Estabrooks, Ph.D, Science Digest April,1971, pp44-50 EPI6028

The "Manchurian Candidate" could be created. In the years since there have been many claims by people that they were used in experiments by our government where their personalities were fractured in this way. It is difficult to prove but what is clear is human victims were created in these various programs.

Estabrooks died in 1973 just before the Congressional Hearings on the CIA's use of mind control were held. He would have likely been one of the experts called upon in the field. His writings during the "cold war" were all very revealing including his rational for why, when and where controlling a person through hypnosis could be justified. Revealing his thoughts in his published works he attempted to rationalize his research with unwitting subjects. He deliberately created multiple personalities, fracturing personalities through induced trauma and then reprograming people. All of these ideas were discussed in two of his most important books, *Hypnotism*[26](1943,revised 1957) and *The Future of the Human Mind*[27] (1961). The idea that there were military uses was revealed in a chapter in *Hypnotism* called "Hypnotism in Warfare" as was his rational for justified research and use of these technologies. He also believed that LSD was useful in researching the human mind. LSD would be later used in mind control programs in the CIA[28].

In another one of Estabrooks books, *Spiritism*[29] (1947), his views on multiple personalities, trace personalities, telepathy and other ESP phenomena are discussed in the context of his experience in mind control through the use of hypnosis. A review of this individual's writing provides an indication of the kind of thinking that has led to the many abuses of these technologies that have been reported.

26. *Hypnotism,* by G.H. Estabrooks, 1943, revised 1957, EPI6084
27. *The Future of the Human Mind,* by G.H. Estabrooks, 1961. EPI6085
28. Report to the President by the Commission on CIA Activities Within the United States, June, 1975. EPI6014
29. 29. *Spiritism,* by G.H. Estabrooks, 1947, EPI6086

Is it ever right to use "unwitting" people as experimental guinea pigs to test new technology without clear informed consent? How many men in Germany, after World War II, were sent to the gallows to hang by their necks for believing that human experimentation was the right thing to do? The country largely responsible for those trials, the United States, would repeat these violations of human rights by experimenting on thousands of citizens without either their informed consent, or, for that matter, even their knowledge that the experiments were occurring.

"During World War II and the Cold War era, DOD and other national security agencies conducted or sponsored extensive radiological, chemical and biological research programs. Precise information on the number of tests, experiments, and participants is not available, and the exact numbers may never be known. However, we have identified hundreds of radiological, chemical, and biological tests and experiments in which thousands of people were used as test subjects. These tests and experiments often involved hazardous substances such as radiation, blister and nerve agents, biological agents, and lysergic acid diethylamide (LSD). In some cases, basic safeguards to protect people were either not in place or not followed. For example, some tests and experiments were conducted in secret; others involved the use of people without their knowledge or consent or their full knowledge of the risks involved."[30]

> "At least 500,000 people were used as subjects in Cold War-era radiation, biological and chemical experiments sponsored by the federal government"[31]

30. The United States General Accounting Office, Human Experimentation: An Overview on Cold War Era Programs. GAO/T-NSIAD-94-266, Sept. 28, 1994. EPI618

31. MacPherson, Karen. "500,000 Endangered by Tests Since 1940." *Washington Times,* Sept. 29, 1994. EPI620

"Government researchers appear to have had a pattern of choosing 'vulnerable' people – minorities, the poor, prisoners and retarded children – for Cold War-era radiation experiments, Energy Secretary Hazel O'Leary said. 'This appearance of treating some citizens as 'expendable' is especially repugnant.'"[32] The groups which were targeted for these experiments were those with the least opportunity to object, the uneducated or those easily victimized. "Children in orphanages were used to test experimental vaccines for diphtheria and whooping cough for several decades after W.W.II,"[33] according to a newspaper story appearing in 1994. "At least 500,000 people were used as subjects in Cold War-era radiation, biological and chemical experiments sponsored by the federal government, a congressional agency said yesterday...the tests conducted from 1940 through 1974, ranged from radiation to biological and chemical agents like mustard gas and LSD."[34]

The Basic Rationale for the Big Lie

Public safety is the usual logic which is applied to extreme violations of civil law by governing authorities. It is their tradeoff for peace. In December, 1998, the United States Secretary of Defense made some interesting observations, observations of the kind that have always formed the basic rationale for repression. He said:

> "The best deterrent that we have against acts of terrorism is to find out who is conspiring, who has the material, where they are getting it, who they are talking to, what are their plans. In order to do that, in order to

32. MacPherson, Karen. "Radiation Researchers Went After 'Vulnerable' Subjects." *Anchorage Daily News*, Jan. 26, 1994. EPI1183
33. AP. "Vaccines Were Tested on Orphans." *Anchorage Daily News*, June 11, 1997. EPI75
34. MacPherson, Karen. "500,000 Endangered by Tests Since '1940." *Washington Times*, Sept. 29, 1994. EPI620

interdict the terrorists before they set off their weapon, you have to have that kind of intelligence-gathering capability, but it runs smack into our constitutional protections of privacy. And it's a tension which will continue to exist in every free society – the reconciliation of the need for liberty and the need for law and order.

And there's going to be a constant balance that we all have to engage in. Because once the bombs go off – this is a personal view, this is not a governmental view of the United States, but it's my personal view – that once these weapons start to be exploded people will say protect us. We're willing to give up some of our liberties and some of our freedoms, but you must protect us. And that is what will lead us into this 21st century, this kind of Constitutional tension of how much protection can we provide and still preserve essential liberties."[35]

Although Secretary of Defense Cohen indicates that this is his personal view and not the view of the U. S. Government, what does this say about the position of the second in command of the most powerful military organization in the world at that time? We believe that this explains how lies can be told without a blink of the eye or the twitch of a muscle. We believe that this explains how people in high positions rationalize their behavior when the very rights of those they are charged to protect are violated.

The Patriot Act is a good example of the overreaching. This Act has been objected to by over a hundred local and state governments. Even with the Act, and special secret

35. Cohen, Secretary of Defense William S. Hemispheric Cooperation In Combating Terrorism, Defense Ministerial of the Americas III. *Defense Viewpoint,* Dec. 1, 1998. EPI660

courts created years ago, the President has still authorized even more reaches beyond the Constitution in the "war on terror". Fear is being elevated and the Constitution is being lost in the shuffle.

The Beam Team

There have been stories that have been made public that describe some of the exotic uses and misuses of these technologies. From the 1960's through the 1970's, there were news and intelligence reports of microwave beams being directed at the U. S. Embassy in Moscow. There are some indications that beaming of the Embassy was also conducted as late as 1983. This beaming has been the subject of much speculation over the years. The government has not released the full story, and for the most part it remains classified.[36, 37]

Speculated effects of the radiation include health impairment and mind manipulation.[38] Beginning in 1965, the government tested embassy personnel for genetic damage, which may have been caused by the microwave beaming. At the same time, they launched Operation Pandora, where monkeys were exposed to the same type of signals in order to measure the effects.

While government representatives have maintained that there were no ill effects from the microwaves, the results of the embassy personnel testing and Pandora remain classified.[39] Even though the impact of Russian microwave radiation is unclear, it turns out that there may very well have been some effect. When Dr. Gottlieb, the MKULTRA program director for the CIA, testified to the United States Congress, he said

36. Cross Currents, The Perils of Electropollution, The Promise of Electromedicine, by Robert O. Becker, M.D.
37. Super-Memory- The Revolution, by Sheila Ostrander and Lynn Schroeder,1991.
38. Ibid.
39. "The Mind Fields", by Kathleen McAuliffe, Omni Magazine, February 1985.

that when President Nixon went to the USSR in 1971, members of his party showed abnormal behavior including crying and depression. The spy community already knew that the Soviets had developed microwave beaming technologies which can affect mind, memory and health. Soviet research had already shown that it was also possible to create hallucinations and significant perceptual changes in people.[40]

The CIA's use of mind altering technologies is not new. Under MKULTRA, the CIA conducted memory experiments on thousands of unknowing people across the United States in 180 hospitals, research centers and prisons. The CIA used LSD and other drugs, brainwashing, sensory deprivation, hypnosis and many other mind control techniques until 1976 when the United States Senate investigated these practices.[41,42,43] It appears the names of the programs and their approaches changed over the years but the programs have continued.[44]

Meanwhile the Soviets had moved way ahead of the United States in their development of mind war technologies. They had perfected a device called the LIDA machine that produced an Extremely Low Frequency (ELF) pulsing field. LIDA was used to put prisoners of war into trance, so that secrets could be extracted from them more readily. As mentioned later, this device was tested by Dr. Ross Adey in the United States at the Loma Linda VA Hospital. The Soviets went on to discover that by slight manipulation – by reversing brain polarity (by passing very low voltage current through the front of the brain to the back) – they could induce deep sleep.

40. Super-Memory- The Revolution, by Sheila Ostrander and Lynn Schroeder,1991, pages. 298-299.
41. Journey into Madness, The True Story of Secret CIA Mind Control and Medical Abuse, by Gordon Thomas, 1989.
42. Super-Memory- The Revolution, by Sheila Ostrander and Lynn Schroeder,1991, pages. 292-304.
43. Report to the President by the Commission on CIA Activities Within The United States, June, 1975, pages 225-231.
44. Journey into Madness, The True Story of Secret CIA Mind Control and Medical Abuse, by Gordon Thomas, 1989.

Mounting evidence also indicated that they may have perfected "telepathic hypnosis" which could be deployed from hundreds of miles away from the target person.[45]

The Soviets took their technologies a huge jump forward by 1975. It was that year they began using seven giant radio transmitters to pulse ELF waves in the 3.26 to 17.54 megaHertz range. These waves were pulsed at 6 and 11 Hertz – key brain wave rhythms – and became known to ham radio operators as the "woodpecker" signal. The Soviet story, like the HAARP story, indicated that these were used for communication with submarines, but many believe that the negative side effects were intentional.[46] These "side effects" have been speculated to have caused communications interference; power failures; mood alterations over significant areas, affecting a large percentage of the population; and weather modifications which have had a devastating effect on food production since the 1970's.[47]

Remote Controlled People: MKULTRA

"Dr. Gottlieb, born August 3, 1918, was the CIA's real-life 'Dr. Strangelove' – a brilliant biochemist who designed and headed MKULTRA, the agency's most far-reaching drug and mind-control program at the height of the Cold War. Though the super-secret MKULTRA was ended in 1964, a streamlined version called MK-SEARCH was continued – with Gottlieb in charge – until 1972."[48] During this period substantial interest in mind control was stimulated by Soviet use of microwaves. In 1988, "thirty-five years after security

45. Super-Memory- The Revolution, by Sheila Ostrander and Lynn Schroeder,1991, pges. 292-304.
46. Super-Memory- The Revolution, by Sheila Ostrander and Lynn Schroeder,1991, pges. 299-302.
47. Ibid.
48. Foster, Sarah. "Cold War Legend Dies at 80: Famed as CIA's Real-life 'Dr. Strangelove.'" *Worldnetdaily,* March 9, 1999. EPI279

officers first noticed that the Soviets were bombarding the U. S. Embassy in Moscow with microwave radiation, the U. S. government still has not determined conclusively – or is unwilling to reveal – the purpose behind the beams."[49] The government did know what was happening. The Soviets had developed methods for disrupting the purposeful thought of humans and was using their knowledge to impact diplomats in the United States embassy in Moscow.

In 1994 a report concerning the MKULTRA program was issued containing the following information:

> "In the 1950's and 60's, the CIA engaged in an extensive program of human experimentation, using drugs, psychological, and other means, in search of techniques to control human behavior for counter-intelligence and covert action purposes.
>
> In 1973, the CIA purposefully destroyed most of the MKULTRA files concerning its research and testing on human behavior. In 1977, the agency uncovered additional MKULTRA files in the budget and fiscal records that were not indexed under the name MKULTRA. These documents detailed over 150 subprojects that the CIA funded in this area, but no evidence was uncovered at that time concerning the use of radiation.
>
> The CIA did investigate the use and effect of microwaves on human beings in response to a Soviet practice of beaming microwaves on the U. S. embassy. The agency determined that this was outside the scope of the Advisory Committee's purview.
>
> ...The Church Committee found some records, but also noted that the practice of MKULTRA at that time was 'to maintain no records of the planning and approval of test

49. Reppert, Barton. "The Zapping of an Embassy: 35 Years Later, The Mystery Lingers." AP, May 22, 1988. EPI1112

programs.' ...MKULTRA itself was technically closed out in 1964, but some of its work was transferred to the Office of Research and Development (ORD) within the DS&T under the name MKSEARCH and continued into the 1970's.

The CIA worked closely with the Army in conducting the LSD experiments. This connection with the Army is significant because MKULTRA began at the same time that Secretary of Defense Wilson issued his 1953 directive to the military services on ethical guidelines for human experiments.

Throughout the course of MKULTRA, the CIA sponsored numerous experiments on unwitting humans. After the death of one such individual (Frank Olson, an army scientist, was given LSD in 1953 and committed suicide a week later), an internal CIA investigation warned about the dangers of such experimentation. The CIA persisted in this practice for at least the next ten years. After the 1963 IG (Inspector Generals) report recommended termination of unwitting testing, Deputy Director for Plans Richard Helms (Who later became Director of Central Intelligence) continued to advocate covert testing on the grounds that 'positive operational capability to use drugs is diminishing, owing to a lack of realistic testing. With increasing knowledge of state of the art, we are less capable of staying up with the Soviet advances in this field.'...Helms attributed the cessation of the unwitting testing to the high risk of embarrassment to the Agency as well as the 'moral problem.' He noted that no better covert situation had been devised than that which had been used, and that 'we have no answer to the moral issue.'"[50]

50. Advisory Committee Staff, Committee on Human Radiation Experiments. Methodological Review of Agency Data Collection Efforts: Initial Report on the Central Intelligence Agency Document Search. June 27, 1994. (HG) EPI579

They did have the answers to the moral questions on human experimentation but chose to ignore them, destroying the records, hiding the truth and still continuing in their efforts. Nothing has changed as each participating organization, using national security laws, avoids disclosure and accountability. The records which were destroyed contained the evidence necessary to perhaps send some administrators to jail for society's version of behavior modification. Once again, there was no accountability and no recognition of the rights of the individuals damaged by these experiments.

MKULTRA had many implications for mind control beyond chemicals in about 150 subprojects. Soft music, voice, flickering light and patterned light were being used as far back as 1956[51] to gain influence over the minds of test subjects and unwitting victims. The idea of remote control of people was also part of the project funded under MKULTRA Subproject No. 94[52]

In a press report in late 2005, Japan's largest telecommunications company, the Nippon Telegraph and Telephone Corporation, announced the invention of a device that could be used for remotely controlling people[53]. This technology, according to the article, could be used in any sound device for enhanced virtual reality or sound performance. The idea of mimicking professional dancers forcing the listener to move as the dancer moves without being able to resist and other innovations were discussed. The thought that a technology can cause involuntary movement should be considered with great care. In combination with other technologies described in this book a great deal could be done; "It's a mesmerizing sensation similar to being drunk or melting into

51. Memorandum for the Record, MKULTRA Subproject 49, February 16, 1956, from Sidney Gottlieb. EPI6005
52. Memorandum for the Record, MKULTRA Subproject 94. EPI6006
53. A Remote Control that Controls Humans, Headset Sends Electricity Through Head, Forcing Wearer to Move, By Yuri Kageyama, MSNBC, October 25, 2005. EPI6033

sleep under the influence of anesthesia. But it's more definitive, as though an invisible hand were reaching inside your brain." This technology makes a person move around in a controlled way by affecting their brain with electrical impulses. They are expecting to use this technology in virtual reality games and dance clubs to teach very precise movements, and perhaps as an added feature for audio equipment.

This area of the technology starts to move toward what the Air Force now calls "Controlled Effects"[54]. The Air Force's research and applications of controlled effects are oriented toward three areas – equipment/hardware, computer software and ultimately the human operator. The operator of the equipment is the target in the ideal situation. To control the operator or interfere with his/her ability to function is what controlled effects are all about.

An article in the U.S. Air Force publication, *AFRL Technology Horizons* states that, "By studying and modeling the human brain and nervous system, the ability to mentally influence or confuse personnel is also possible. Through sensory deception, it may be possible to create synthetic images, or holograms, to confuse an individual's visual sense or, in a similar manner, confuse his sense of sound, taste, touch or smell."[55] Think about this for a moment – synthetic memory, synthetic experiences that cannot be sorted out from those that are real. This is the stuff of controlled effects.

In a open request for proposal (RFP) the Air Force Research Laboratory was seeking "research in support of the Directed Energy Bioeffects Division of the Human

54.Controlled Effects, The Future of Controlled Effects, By Dr. William Baker (Chief Scientist) and Dr. Eugene J. Beddnarz of the Air Force Research Laboratory's Directed Energy Directorate and Dr. Robert L. Sierakowski, *AFRL Technology Horizons*, June, 2004, EPI6034
55. Ibid.

Effectiveness Directorate."[56]The RFP stated the "ultimate goal of such research is to develop a fully articulated theory, with supporting predictive models that will facilitate the inducement of desired behavioral effects in individuals and groups through the use of Nonlethal Weapons." This contract request will remain open for project submitals through 2009.

Military sponsored organizations meet regularly in order to foster innovation in these merging areas of research. The Directed Energy Professional Society, is basically a thinktank of the United States military and their surrogates. Navy, Air Force, military contractors and other invited persons can participate. All are required pre-conference security clearing and are subject to restrictions on what sessions they may attend. Using "professional societies" has always been a way to mask military research but today, because of the powerful nature of the research, it is contained in only being accessible to those with a "need to know". Conferences are organized by this group in many areas including those that impact human biology and new weapon systems such as those discussed throughout these pages. The group encourages integration and crossovers into other scientific disciplines[57] in the hope that it will stimulate new innovation and reach the organizations ultimate objectives as they have expressed them in their publicly released materials. These conferences including the Ninth Annual Directed Energy Symposium, scheduled for October 30- November 3, 2006, are heavily dominated by military personnel seeking "controlled effects" applications and other uses.[58]

56. Air Force Research Laboratory, "Research in Support of the Directed Energy Bioeffects Division of the Human Effectiveness Directortate", BAA 05-05 HE, Open until September 30, 2009. EPI6068
57. "From Technology Trenches", Major General Donald L. Lamberson (USAF Ret.), Opening Remarks Letter, 2006, EPI6081
58. "Directed Energy Professional Society" News and Announcements. http://www.deps.org/ EPI6080

Going Back to the history of controlled effects, it was reported in 1968, that Dr. William D. Neff at the Indiana University Foundation had already investigated the code for how sound is transferred within the nervous system. He had the idea of creating a system for injecting sound directly into the brain with the goal of aiding the deaf and for other uses[59]. This was the focus of his work.

In recent years the Defense Advanced Research Projects Agency (DARPA) has pursued research into brain decoding and the development of electronic micro and nanocircuits that will directly interact with the brain. A great deal of material has been released to the public regarding these technologies. There have been breakthroughs in sound, taste, touch and visual systems by integrating several areas of science. DARPA has assembled an array of specialists whose cooperation is yielding big payoffs and rapid advances in this area.[60] In addition to the overview materials were a series of reports that came out of a meeting sponsored by DARPA in November, 2000.[61] This conference, now over six years ago, indicated that breakthroughs had already been made in decoding the brain. The reports also contained a great deal of information on the state of implant technology as it relates to the brains of humans and animals. New microchip implant technology could be used for direct interaction between the brains of people and computers.

One of the institutions cooperating in this research is the University of Arizona. Interestingly, a few days after the DARPA conference, *New Scientist* reported[62] that the university had succeeded in implanting the brains of monkeys and

59. BioScience "Capsule", number 18, September 1968, Biological Sciences Communications Project which was sponsored by the United States government. EPI6049

60. http//www.darpa.mil/dso/thrust/biosci/bim/overview.html, Bio Info MICRO Program, Program Overview.EPI6040

61. http://www.darpa.mil/dso/thrust/biosci/bim/briefings.html Bio: Info: Micro Program Kickoff Meeting, Washington, D.C., November 1-2, 2000. EPI6041

62. Power of Thought, by Helen Phillips, *New Scientist,* November 9, 2000. EPI6042

then training them to manipulate a computer screen just by using their own thoughts. At that time the article indicated that the current system that monitors and interprets the brain's information is too bulky to be portable. With the advances in nanotechnology and microcircuitry this is also changing[63] as faster and more portable equipment is produced. In a 2003 report by *New Scientist,*[64] additional advances were noted. They reported that the part of the brain responsible for routing and storing memories – the hippocampus – could be created on a silicon chip. The scientists created a complete model of this part of the brain in rats and "then programmed the model onto a chip" which replaces the natural part of the brain with the artificial one. In order to accomplish this, according to the article, the researchers had to overcome three major hurdles. "They had to devise a mathematical model of how the hippocampus performs under all possible conditions, build that model into a silicon chip, and then interface the chip with the brain." By the fall of 2003, they had advanced this technology to the point were they where testing it on monkeys rather than rats.[65]At that time, they had succeeded in using signals from the monkey's brain to control the movement of a robotic arm. Higher-end emerging technologies such as Field Programmable Gate Arrays (FPGAs) and artificial neural networks (ANNs) have the potential to make these advances orders of magnitudes more powerful and adaptive. As extraordinary as it may seem, these experiments have only begun to scratch the surface. The scientists involved in this work see this as a way to replace damaged portions of the brain allowing for eventually not only the control of robotic arms but actually sending signals to the human body to cause it to respond in specifically directed ways. This science will continue to

63. Multichannel Brain-Signal-Amplifying and Digitizing System, Lyndon B. Johnson Space Center, Houston, Texas, USA. NASA Tech Briefs, December, 2005. EPI6031
64. World's First Brain Prosthesis Revealed, by Duncan Graham-Rowe, *New Scientist,* March 12, 2003. EPI6043
65. Monkey's Brain Signals Control 'Third Arm', by Duncan Graham-Rowe, *New Scientist,* October 13, 2003. EPI6044

advance and although there is concern for the misuses of this technology there is a great deal of promise for those with brain injuries who could benefit by these kinds of advances.

Hypersonic Sound

In another breakthrough, Hypersonic Sound (HSS) was created by inventor Elwood (Woody) Norris. He won the Lemelson-MIT Prize of $500,000 for this invention[66] in 2005. The device allows sound to be transferred through the air in a narrow beam so that only a person in the way of the beam can hear the sounds. No one else around them can hear it, only the targeted person who reports things like "voices in their head." The military adopted a version of the device known as Long Range Acoustic Device (LRAD) which can be used up to 500 yards away as a warning system or can be used as a Nonlethal weapon by cranking the volume up to 120 decibels, "loud enough to disable enemy combatants."[67] Mr. Elwood actually perfected this technology in 2002, according to an ABC News report[68] which said that this invention would produce sound at not 120 decibels, but rather 145 decibels, which is fifty times the human threshold for pain.

About the same time in 2005 *New Scientist* announced a new Sony invention. "The technique suggested in the patent is entirely noninvasive. It describes a device that fires pulses of ultrasound at the head to modify firing patterns in targeted parts of the brain, creating 'sensory experiences' ranging from moving images to tastes and sounds. This could give blind or

66. Inventor Creates Soundless Sound System, Soundless Sound Invention Uses Hypersonic Beam to Produce Audio Seemingly From Inside Listener's Head, by Typh Tucker, Associated Press, Portland, Oregon, U.S.A., April 22, 2005. EPI6036
67. http://web.mit.edu/newsoffice/2005/lemelson-norris.html, Massachusetts Institute of Technology, News Office, Inventor Earns Lemelson-MIT Prize for Sound Thinking, April 18,2005. EPI6037
68. Sound and Fury - Sonic Bullets to be Acoustic Weapon of the Future, By Judy Muller, July 16, 2002. EPI6045

deaf people the chance to see or hear, the patent claims."[69] This gets to the heart again of "controlled effects" being actively sought by researchers and the military with very different objectives in mind.

The Foreign Broadcast Information Service, a program operated by the CIA, translates media reports, papers and other published data that it makes available to policy makers and others for their use. In one release Chinese efforts in "controlled effects" were noted. The topic of the paper was "Health, Military, Proliferation, Technology"[70]. This paper also puts new energy-based weapons technology into the same groupings as the U. S. Air Force for their "controlled effects". The Chinese report on infrasound weapons such as those developed by Elwood and others. The article states, "A small amount of output power can induce immeasurable fear and cause mass hysteria. A large amount of output power can cause unstable mental states and body malfunction, or even symptoms of mental disease." This is through the use of infrasound – sound outside of the range of normal hearing. A number of other capabilities were also pointed out in that paper unrelated to the topics covered in this book which deal with the whole array of energy-based weapon systems[71].

Back to MKULTRA

The evolution and history of invention, according to corporate records, patents and other reports shows that many new technologies have been combined to provide, in the 21st century, the state of the art advances that are now available. These early projects began to form the basis of the concepts that now have advanced for over eighty years.

69. Sony Patent Takes First Step Towards Real-life Matrix, By Barry Fox and Jenny Hogan, New Scientist, April 7, 2005. EPI6038
70. Foreign Broadcast Information Service, Item Number 00617177, Source: China , Beijing Renmin Junyi People's Military Surgeon In Chinese, Vol. 40 No. 9 Sept., 1997 pp. 507-508. EPI3381
71. *Earth Rising - The Revolution: Toward a Thousand Years of Peace,* By Begich & Manning, Sept., 1995, Earthpulse Press. (650 sources quoted). EPI6054

A complete understanding of what could effect the body and mind was the basis of the earlier work. MKULTRA Subproject No. 99 was a project that sought to "support studies on the optical rotatory power of solid and liquid crystals" for their possible application in developing "practical methods for modulation of light intensity by electrical fields, and the development of simple optical shutters." The intention of the study was to use the information to enhance physical studies of methods which will influence the human nervous system.[72]

MKULTRA Subproject 119 involved "a critical review of the literature and scientific developments related to the recording, analysis and interpretation of bioelectric signals from the human organism, and activation of human behavior by remote means."[73] The project was designed to address five main areas of interest:

a. Bioelectric sensors: sources of significant electrical potential and methods for pick up.

b. Recording: amplification, electronic tape and other multichannel recording.

c. Analysis: autocorrelators, spectrum analyzers, etc., and coordination with automatic data processing equipment.

d. Standardization of data for correlation with biochemical, physiological and behavioral indices.

e. Techniques for activation of the human organism by remote electronic means.[74]

72. Memorandum for the Record, MKULTRA Subproject 99, Sept.1, 1961. EPI6007
73. Memorandum for the Record, MKULTRA Subproject 119, August 17, 1960. EPI6008
74. Ibid.

In a 1976 United States patent it appears that it was possible to monitor brain activity and then alter it.[75] Although the apparatus was cumbersome, it was an early model of what was possible even over thirty years ago. The idea was to review, interpret and then control brain activity using a radio frequency signal.[76] More recently much more powerful systems have been developed for looking at the brain in real time on a computer screen. These new devices use brain biofeedback in combination with other signal generators to alter a person's brain activity and patterns.

In a 1963 memorandum, marked "Eyes Only" from the Deputy Director for Plans, Richard Helms to the Deputy Director of Central Intelligence[77], much is revealed; Helms makes the case for the use of "unwitting" test subjects. The memo states that "for over a decade the Clandestine Services has had the mission of maintaining a capability for influencing human behavior; and testing arrangements in furtherance of this mission should be as operationally realistic and yet as controllable as possible." The memo indicates an ongoing relationship with a number of police departments in principal U. S. cities, the Bureau of Narcotics, Bureau of Prisons, the Department of Justice and "various Foreign intelligence and/or security organizations having current and continuing interrogation problems."[78]It is interesting that the same thing happening in 1963 is being repeated in 2005-2006 with the transportation of individuals of interest to the United States to other third world countries or locations, where basic human rights are

75. United States Patent Number 3,951,134, Apparatus and Method for Remotely Monitoring and Altering Brain Waves, April 20, 1976, Inventor Robert Malech. EPI6011
76. Multichannel Brain-Signal-Amplifying and Digitizing System, Lyndon B. Johnson Space Center, Houston, Texas, USA. NASA Tech Briefs, December, 2005. EPI6031
77 .Memorandum from the Deputy Direct for Plans, Richard Helms to the Deputy Director of Central Intelligence,Subject: Testing of Psychochemicals and Related Materials, December 7, 1963. EPI6009
78. Memorandum from the Deputy Direct for Plans, Richard Helms to the Deputy Director of Central Intelligence,Subject: Testing of Psychochemicals and Related Materials, December 17, 1963. EPI6009

ignored, and the most brutal and invasive interrogation methods applied. Helms, for those who do not remember, later became the Director of Central Intelligence (1966-1973). This was during the Nixon years where he served until the Watergate scandal drove that administration out of office. The CIA was also in trouble, along with the FBI, for their abuses of power which were then under investigation because of their activities within the United States. This included the CIA's MKULTRA and related projects that were the subject of a Presidential Commission inquiry and report issued in 1975.[79] This 300 page report revealed part of the underlying corruption of the Nixon Administration and its predecessors. The report addressed the use of mind control and other abuses in violation of United States law including matters of privacy, mail intercepts, and invading the ranks of various groups opposed to United States policy as it related to the political agenda of the Executive Branch and more. During the time of the investigation the agency was under the authority of the new Director of Central Intelligence, George H. Bush, later to become the 41st President of the United States.

Politics and scandal seem to cycle through every few years. The players change but the outcomes are always rooted in someone else's "good intentions" in trying to do something "for our own good" not quite working out. The level of corruption in 1973, when the cards came tumbling down for Nixon, had not been matched until recent days under the administration of George W. Bush, the 43rd President of the United States and his Vice President, Dick Cheney. We are back to where we were before those CIA inquiries of the 1970's. Now with the power of far superior modern technology, privacy issues have become the issues of the 21st Century, with controlling the human mind a very significant matter. What are leaders willing to do to further their own

79. Report to the President by the Commission on CIA Activities Within the United States, June, 1975. EPI6014

ends, which they believe are for our own good? What things will be lost in the fear created by the latest, "war on something, almost anything" mentality, that makes the average person give up their personal liberty in fear?

Hearing without Ears

The answers start with the understanding of the illusions that could be created with the misuse of technology or the proper uses that these new concepts offer. There have been several researchers who have made it possible to hear without the use of the ears or normal hearing pathways to the brain. In one U. S. patent "A method and apparatus for the simulation of hearing in mammals by introduction of a plurality of microwaves in the region of the auditory cortex is shown and described. A microphone is used to transform sound signals into electrical signals which are in turn analyzed and processed to provide controls for generating a plurality of microwave signals at different frequencies."[80] These signals are then applied to the head of the subject so that the auditory cortex in the brain is affected and sound is heard in the head without the interaction of the ears or normal auditory pathways through the inner ear and the eighth cranial nerve. Microwave hearing was first observed and reported by Dr. Allan H. Frey[81] using microwave (radio frequency energy or RF) to induce hearing in deaf individuals and other people who had normal hearing. Pulse-modulating a microwave to carry a voice or other information was now possible. In a later invention by Philip L. Stocklin, Patent Number 4,858,612, Stocklin created an advanced hearing device. This was one of many design advances created in the 1980's for hearing without the "normal" hearing mechanisms being involved.

80. United States Patent 4,858,612, Hearing Device, August 22, 1989, Inventor: Philip L. Stocklin EPI6010
81. Human Auditory System Response to Modulated Electromagnetic Energy, Allan H. Frey, General Electric Advanced Electronics Center, Cornell University, Ithaca, NY, USA, Applied Physiology 17(4):689-692, 1962. EPI6015

In a paper appearing in *Biological Effects and Health Implications of Microwave Radiation*[82] was a discussion of the health effects of microwave energy. That included Dr. Frey, who said, "...I have come to the conclusion that there is something happening there...that there is a headache type of phenomena. I can see ways and means experimentally to explore this. I have not done so with humans because I think there is an ethical question here: I have seen too much. I very carefully avoid exposure myself and have for quite some time now. I do not feel that I can take people into these fields and expose them in all honesty and indicate to them that they are going into something safe."

This is a very important observation from one of the early researchers into the hearing effects of microwaves. Dr. Frey's cautions follow the "precautionary principle" applied in Europe when considering the health implications of all new technology. This principle simply states that when the information is emerging, and safety implications are noted, people are informed so that they can make decisions based on the best information available rather than waiting for detailed proofs and explanations.

Dr. Frey had also completed a paper on the biological function of modulated radio frequency energy.[83] The paper was an early review of the literature and had cautionary notes throughout the document. What it addressed were modulated signals that deal with the frequency of the signal. Later research by others determined that pulse-modulated signals were in fact even more powerful. For example, by 1989, a U. S.

82. *Biological Effects and Health Implications of Microwave Radiation,* Maximum Admissible Values of HF and UHF Electromagnetic Radiation at Work Places in Czechoslovakia, by Karel Marha, Institute of Industrial Hygiene, and Occupational Diseases, Prague, Czechoslovakia, Undated. EPI6018

83. Biological Function and Influence by Low-Power Modulated RF Energy, By Dr. Allan H. Frey, Senior Member IEEE, IEEE Transactions on Microwave Theory and Techniques, Vol. MTT-19, No. 2, February, 1971. EPI6050

patent was filed which dealt with the effect of pulse-modulated signals.[84] Through the use of such a signal, hearing could be stimulated in the person whose head was irradiated with microwave signals. The result was the "voice in the head" effect. In a later patents by Hendricus G. Loos[85 86], researchers showed that by pulsing a signal one could override the nervous system and produce relaxation, drowsiness or sexual excitement depending on how the technology was applied.

Loos' patent said, "Physiological effects have been observed in a human subject in response to stimulation of the skin with weak electromagnetic fields that are pulsed with certain frequencies near 1/2 Hz or 2.4 Hz, such as to excite a sensory resonance." Many computer monitors and TV tubes, when displaying pulsed images, emit pulsed electromagnetic fields of sufficient amplitudes to cause such excitation. It is therefore possible to manipulate the nervous system of a subject by pulsing images displayed on a nearby computer monitor or TV set. For the latter, the image pulsing may be imbedded in the program material, or it may be overlaid by modulating a video stream, either as an RF signal or as a video signal. The image displayed on a computer monitor may be pulsed effectively by a simple computer program. For certain monitors, pulsed electromagnetic fields capable of exciting sensory resonances in nearby subjects may be generated even as the displayed images are pulsed with subliminal intensity."[87] Some of the technology was designed to be operated remotely or be delivered on other carriers and can subtly override our natural normal mental or physical states.

84. United States Patent No. 4,877,027, Hearing System, October 31, 1989, Inventor Wayne B.Brunkan. EPI6051
85. United States Patent No. 6,017,302, Subliminal Acoustic Manipulation of Nervous Systems, January 25, 2000, Inventor Hendricus G.Loos. EPI6052
86. United States Patent No. 6,506,148,Nervous System Manipulation by Electromagnetic Fields from Monitors, January 14, 2003, Inventor Hendricus G.Loos. EPI6069
87. Ibid.

These patents are representative of some of the patents that have made it into the public domain where a record of their development could be seen and considered. Covert or other abuses of these technologies was raised as an issue in *Scientific American* in July, 2003[88].

Manipulation of Emotion and Mind

The science of affecting emotional states and other aspects of consciousness was advancing in the private sector as well. A number of breakthroughs were being made in the 1980's and 1990's for enhancing human performance.

Robert Monroe filed two patents in the early 1990's[89] [90] that were intended to be used by average people for enhancing individual human performance. Monroe developed a series of audio materials (audio CD's) for use in behavior modification. He created, through a combination of binaural beat (when two different sound signals are introduced one in each ear) and imposed average EEG signals through stereo headsets as a way to cause a frequency following response (FFR). FFR is where the brain begins to follow the pulse rate of the device that is creating the signal. What was found was that by looking at the average brain activity of a number of individuals who could exhibit very specific brain states, their brain signals could be imposed on others and their effect amplified by binaural beat.

88. "You Can Patent that? A selection of Recently Issued Intellectual-Property Gems", by Gary Stix, *Scientific American*, July, 2003. EPI 6070
89. United States Patent 5,213,562, Method of Inducing Mental, Emotional and Physical States of Consciousness, Including Specific Mental Activity in Human Beings, May 25, 1993, Inventor Robert Monroe. EPI6012
90. United States Patent 5,356,368 Method of and Apparatus for Inducing States of Consciousness, October 18, 1994, Inventor Robert Monroe. EPI6013

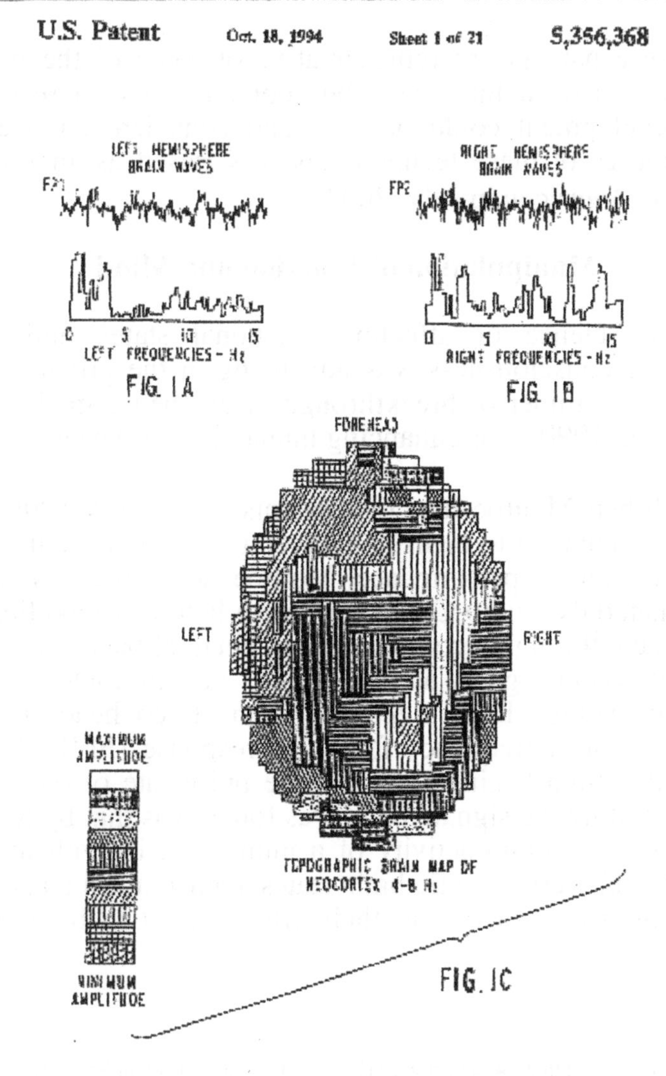

EEG graphic before the use of Hemi-sync®

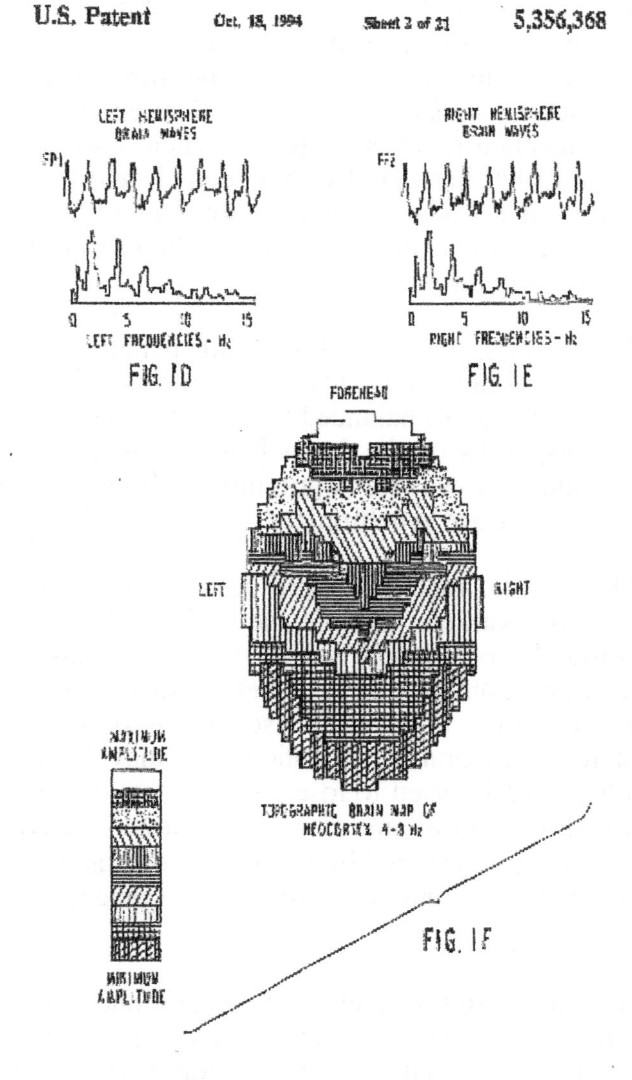

EEG graphic while Hemi-sync® is being used with a stereo headset.

Use of sound for inducing brain states is a bit tricky because we cannot hear below a certain frequency. These lower frequencies are created inside the head when two different sound signals are introduced in each ear – binaural beat. A cancelation effect is created when a sound enters one ear at say 15000 hertz (pulses per second) and the other ear at 15007 leaving a "beat" frequency of 7 hertz, as an example. When the beat frequency is created the brain follows the signal mimicking its pulse rate.

In this example, the whole brain is activated at 7 hz. A look at the brain indicates that in this state, energy is distributed in a more balanced way on both sides of the brain. Both hemispheres of the brain begin to work together in a manner that allows for optimum performance of mental functions – they are synchronized.

Other audio materials were developed by Monroe based on these observations, and the use of the technology with tens of thousands of people over the years has proven its effectiveness. Some of these materials include not just sound signals but voice as well. These are used to anchor ideas related to behavioral modification that may be desired. The technology can be used to lose weight, stop smoking, improve meditation, relaxation and many other uses. The *Hemi-sync*™ systems[91] were developed by Monroe as outstanding tools for self-improvement and increasing our general capacity as people.

In United States Patent #4,889,526[92] another method and apparatus was developed to reduce pain using pulsed electromagnetic signals that cause ion flow in the nervous system in a manner that reduces the perception of pain in the

91. See information on Hemi-sync ™. www.earthpulse.com
92. United States Patent No. 4,889,526, Non-Invasive Method and Apparatus for Modulating Brain Signals Through an External Magnetic or Electric Field To Reduce Pain, December 26, 1989, Inventors: Elizabeth Rauscher and William Van Bise. EPI6016

human body. The creators say that, "This invention relates to an electronic apparatus which is capable of generating a magnetic field that is precisely tuned in order to interact with the brain and heart in order to pace the heart and also interact with the nervous system in order to counteract pain." William Bise, the inventor of this technology, was interested in looking at the effects of low levels of energy on people. He had published a 1978 article[93] where he concluded "that very low power-density RF can interfere with the normal brain-wave patterns of humans; and if these patterns are meaningful in terms of mood and behavior, then RF affects these parameters as well." This is again one of the early calls for caution in this area of science during a time of greatly increasing levels of radio frequency (RF) energy being added to the environment by new communications systems used throughout the world.

An additional patent that was found to be particularly interesting concerns the use of either very low or very high frequency audio signals for delivering subliminal information.[94] This invention was intended "(a) to provide a technique for producing a subliminal presentation which is inaudible to the listener(s), yet is perceived and demodulated (decoded) by the ear for use by the subconscious mind. (b) to provide a technique for transmitting inaudible subliminal information to the listener(s) at a constant, high level of signal strength and on a clear band of frequencies. (c) to provide a technique for producing inaudible subliminal presentations to which music or other 'foreground' programming may be added, if desired." This is a very important patent when the implications are considered. This system bypasses the conscious mind and drops the information into the subconscious in a manner that avoids any conscious filtering of the information. The

93. *Low Power Radio-Frequency and Microwave Effects on Human Electroencephalogram and Behavior,* Physiology Chemistry & Physics 10 (1978) Pages 387-398. EPI6019

94. United States Patent No. 5,159,703, Silent Subliminal Presentation System, October 22, 1992, Inventor Oliver Lowery. EPI6017

information, without conscious review may, in fact, set up conflicts in the listener's belief systems. In other words the messages may not line up with the way the individual person thinks, which can lead to a number of problems. Conflicts in belief systems can lead to significant emotional and psychological problems, which is why when "subliminals" are used a person should know exactly what is being said to assure that the messages line up with the beliefs of that individual. A person should be able to hear or review all of the words being used to reshape beliefs or behaviors when using these kinds of technology.

Light and sound systems like the *Sirius* or *Proteus* devices also allow for the input of a users own recorded voice for the maximum effectiveness when the brain is otherwise being moved into an altered state by FFR. Connecting a recording through these devices allows for self-programming in creating the optimum state of awareness while utilizing the whole brain effect.

There are several methods that are being developed for changing the way the brain operates. These tools are being created and they will be used to continue to enhance the human experience or degrade that experience. The brain, the master control of body and mind, is the target of these technologies the impacts of which will be determined by the operators.

Chapter Three

New Initiatives

Lieutenant Colonel John B. Alexander

From his offices at Los Alamos Labs, John Alexander was one of the prime movers in the advanced development of nonlethal weapons systems. He pursued his interests in obscure science and parapsychology, connecting with Janet Morris, with whom he wrote a book on mind training techniques.[95] In putting the book together, Janet recruited Ray Cline, a former deputy director of the CIA, who opened doors to the White House and Pentagon, according to a Wall Street Journal article.[96] The use of the technologies was apparently known by Alexander to be problematic, because some of the weapon systems would violate international agreements. Moreover, individuals began to raise concerns suggesting that the use of "nonlethals" might cause escalation, rather than control, of volatile conflicts.

The Pentagon's nonlethal study group concluded that a major effort should be made to develop these technologies, and

95. The Wall Street Journal, "Nonlethal Arms, New Class of Weapons Could Incapacitate Foe Yet Limit Casualties, by Thomas E. Ricks, January 4, 1993, page A1 and A4.
96. Ibid.

suggested that President Bush (41st President) announce a new initiative in this area similar to President Reagan's announcement of the Strategic Defense Initiative ("Star Wars").[97]

Military officials objected that such a "new initiative" announcement might have the effect of spurring other governments and possible adversaries to develop their own new systems and that it might cause policy makers to begin "political meddling" into this militarily promising area. These military objections drove the program into even more layers of secrecy, and Janet Morris and Ray Cline were cut out of the inner circle. The policy makers then decreed that the nonlethal technologies would be referred to in softer terms such as "disabling systems" in the future.[98]

The new technology race was already on, and it appeared America's "adversaries" may have already made significant advances in the development of new weapon systems. The limits for microwaves were set in the United States 1,000 times higher than the level considered safe in the former Soviet Union. The reason that the Soviets set their safety standard as low as they did was because they detected biological effects at levels ignored by the West.[99] [100] The Russians abandoned the thermal model to which the U. S. Military ascribed when setting American standards in the 1950's (standards were subsequently adopted by civilian regulators).

97. Ibid.
98. The Wall Street Journal, "Nonlethal Arms, New Class of Weapons Could Incapacitate Foe Yet Limit Casualties, by Thomas E. Ricks, January 4, 1993, page A1 and A4.
99. "The Mind Fields", by Kathleen McAuliffe, Omni Magazine, Feb., 1985.
100. Low-Intensity Conflict and Modern Technology, Lt Col. David J. Dean USAF Editor, Air University Press, Center for Aerospace Doctrine, Research, and Education, Maxwell Air Force Base, Alabama, June, 1986.

The thermal model only acknowledges heating effects as potentially dangerous and disregards the lower level radiations which also have significant effects when interacting with the energy fields of living things. The Soviets showed that electromagnetic fields, well within the U. S. "safety" standards, could disrupt heart rhythms, blood pressure and metabolism. In addition, Soviet scientist A.S. Presman said these fields can produce "visual, acoustic, and tactile sensations in man as well as emotional states in animals, (inducing everything) from suppressed states similar to narcosis to excited states reaching epilepsy."[101] Presman went on to discuss the pronounced negative impact on creatures, from the embryonic stage up to the beginning of sexual maturity. The problem was the idea that genetic alteration could be caused by low-level electromagnetic fields causing deformation, death or other debilitating effects. The lack of mainstream medical understanding, beyond that found in leading edge research facilities, was the underlying barrier to recognizing the risks. What all of this means is that real damage to living things, including people, is likely being done by use of various kinds of energy sources. On the other hand, if we gained a better understanding of these non-thermal levels of energy, positive uses could be more widely applied.

There have been several people over the years who have addressed some of these issues, including Lieutenant Colonel John B. Alexander, U. S. Army, who has written on these topics for over twenty-five years. In a 1980 article[102] he said, "Mind-altering techniques designed to impact opponents are well-advanced. The procedures employed include manipulation of human behavior through the use of psychological weapons affecting sight, sound, smell, temperature, electromagnetic energy or sensory deprivation." The other area that he pointed

101. "The Mind Fields", by Kathleen McAuliffe, Omni Magazine, February, 1985.
102. *The New Mental Battlefield: "Beam Me Up, Spock"*, by Lieutenant Colonel John B. Alexander, US Army,Military Review, December ,1980. EPI6022

to was the area of out-of-body experiences for intelligence gathering. This is the concept referred to today as "remote viewing" – seeing objects and events far removed from the physical location of the "traveler" who "sees" what would otherwise be impossible to perceive. Alexander further stated that diseased conditions could be transferred from one organism to another over distances using only the diseased energetic state as the carrier of the disease; in other words, transmission of illness, without physical contact, which would have no apparent "normal" cause. The idea of interacting with the bioenergetic fields of the human body were also discussed in the article as a way of observing emotional and physical conditions in people and then using this knowledge to manipulate these conditions by imposing energetic signals of very specific form and frequency for affecting the individual. One of the most startling assertions he made was that it may be possible to induce hypnotic states in people from over 1000 kilometers away – and this was 1980!

One of the other areas Alexander discussed was the use of mind-to-mind thought induction techniques (mental telepathy) organized and directed at specific individuals or groups with the recipient(s) of these artificial thoughts unable to determine that they originated from an outside source. The person would believe that the thoughts were their own. These are all important points in the development of the technology because of when these assertions were being made. Now in 2006, a great deal more is possible.

There have been several attempts over the years to get this issue addressed in open scientific conferences and other forums as a human rights issue. Alan Scheflin, a professor of law at the University of Santa Clara, prepared an outstanding paper in 1982, and came to a number of conclusions including the following:

"First, research designed to explicitly control the thoughts and conduct of free citizens is now not only a reality but the evidence is clear that this research is growing in scope, intensity and financing.

Second, frightening progress has already been made by brain behavior researchers in their pursuit of conquest of the mind..."

Professor Scheflin also points out that funding was significantly increasing in 1982, and that there were no international agreements or treaties that dealt with the fundamental human right – *freedom of the mind.* In the paper he traces the history of the CIA's interest in the work of mind control, or what Allen Dulles, the former Director of Central Intelligence in the early 1950's, called "brain warfare". Dulles believed that the Soviet Union was pursuing the goal of controlling the human mind "so that it no longer reacts on a freewill or rational basis but responds to impulses implanted from outside."[103]

One of the leading publications in the world, the *Economist*[104], actually took up the subject of ethics as their cover story. In May, 2002; "Mind Control" appeared in that issue. This lead story started the debate on the ethics of mind control; the ethics of artificially manipulating the human brain for military and other uses.

European Parliament Raises the Issue

We began to address the mind control issue in 1995, in the context of the HAARP public disclosure efforts we were involved with at that time. After several meetings and a public

103. *Freedom of the Mind as an International Human Rights Issue,* by Alan Scheflin, Human Rights Law Journal Volume 3, No1-4, 1982. EPI6023
104. The Future of Mind Control, *The Economist,* May 25-31, 2002, Page 11. EPI6035

hearing sponsored by the European Parliament we were successful in getting the following language included in one of the most comprehensive resolutions on disarmament ever passed by the European Parliament – the relevant section:

> "Calls for an international convention introducing a global ban on all developments and deployments of weapons which might enable any form of manipulation of human beings."[105]

This represents a starting point for the activation of the discussion in Europe and elsewhere on this subject in terms of the new technologies. At the hearing leading to the resolution, I demonstrated an infrasound device. The device allowed people to "hear a voice in their head" while using the machine connected only to a pair of small metal plates about three centimeters in diameter.

I delivered a lecture on Arctic issues and HAARP in Brussels in the European Parliament, on May 5-7, 1997, at the 12th General Assembly Globe International. This was one of the above meetings that led to the above resolution. In attendance were several members of the Russian Duma including Vitaliy Sevastyanov, one of the signers of a resolution by the Russian Duma – their legislative branch of government. Our exposé on HAARP, published in September, 1995, launched the international investigation into the issues surrounding HAARP, including mind control. The resolution that the Russian's passed addressed even the mind altering aspects of the technology in addition to the environmental issues. The following story was released by the Interfax news agency in Russia on August 9, 2002:

105. European Parliament. Resolution on the Environment, Security and Foreign Policy. A4-0005/99. Jan. 28, 1999. EPI159

***Economist*[106] Cover story – May, 2002**

106. The Future of Mind Control, *The Economist,* May 25-31, 2002, Page 11. EPI6035

"MOSCOW (Interfax) - The Russian State Duma has expressed concern about the USA's programme to develop a qualitatively new type of weapon.

Under the High Frequency Active Auroral Research Programme (HAARP), the USA is creating new integral geophysical weapons that may influence the near-Earth medium with high-frequency radio waves, the State Duma said in an appeal circulated on Thursday [8 August].

The significance of this qualitative leap could be compared to the transition from cold steel to firearms, or from conventional weapons to nuclear weapons. This new type of weapons differs from previous types in that the near-Earth medium becomes at once an object of direct influence and its component. These conclusions were made by the commission of the State Duma's International Affairs and Defense Committees, the statement reads.

The committees reported that the USA is planning to test three facilities of this kind. One of them is located on the military testing ground in Alaska and its full-scale tests are to begin in early 2003. The second one is in Greenland and the third one in Norway...

The USA plans to carry out large-scale scientific experiments under the HAARP programme, and not controlled by the global community, will create weapons capable of breaking radio communication lines and equipment installed on spaceships and rockets, provoke serious accidents in electricity networks and in oil and gas pipelines and ***have a negative impact on the mental health of people populating entire regions*** (emphasis added), the deputies said.

They demanded that an international ban be put on such large-scale geophysical experiments. The appeal, signed by 90 deputies, has been sent to President Vladimir Putin, to the UN and other international organizations, to the parliaments and leaders of the UN member countries, to the scientific public and to mass media outlets. Among those who signed the appeal are Tatyana Astrakhankina, Nikolay Kharitonov, Yegor Ligachev, Sergey Reshulskiy, *Vitaliy Sevastyanov*, Viktor Cherepkov, Valentin Zorkaltsev and Aleksey Mitrofanov."

Although these actions are now several years old there appears to have been no further action on these matters by either the European Parliament, the Russian Duma or the United States.

Mankind Research Unlimited, Inc.

One of the more interesting companies involved in psychic and mind control related research in the 1960's-1980's was Mankind Research Unlimited, Inc. (MRU). This company and its affiliates were researching psychic warfare,bioenergetic fields and the manipulation of energy in order to affect people. After reading a *Covert Action*[107] article I decided to take a closer look at the company considering that two of the authors I had published were affiliated in some way with them it seemed a little closer look was in order. In my search a number of things were found that caused me some concern in terms of who was using the research products of MRU and what they were doing with the information.

The President of the company disavowed any affiliation with the CIA but the materials I found on the internet indicated that if this was the case, then the CIA certainly shared the

107. Mind Control: The Story of Mankind Research Unlimited, Inc., by A.J. Weberman,Covert Action Number 9 (June 1980), pges 15-21. EPI6024

same interests and required the same research products as MRU. I have also had occasion to get to know the former Northwest Field Director for the organization, who was involved in much of the early research. Organizations like Mankind Research Unlimited Inc., and its affiliated organizations, which included Systems Consultants Incorporated and the Mankind Research Foundation Incorporated, could easily unwittingly, or through the independent actions of those affiliated with them have cooperated with CIA projects. At the same time there were many good people being used by government agencies without their knowledge. We may never know who knew and who didn't. The following was found describing some of MRU's positively oriented capabilities:

> "LISTING OF HEALTH AND WELFARE PROGRAMS WHICH ARE WITHIN THE CAPABILITY OF MANKIND RESEARCH UNLIMITED, INC.
>
> 1. Investigations into Causal and Preventive Factors for the Sudden Infant Death Syndrome (SIDS)
>
> 2. Methods of Improving Beds, from a Therapeutic and Comfort Standpoint, for Hospital Patients
>
> 3. Applications of Music-Color and Chromotherapy as Remedial Agents in the Treatment of Emotionally Disturbed and Retarded Children
>
> 4. Analysis of the French-Developed Electronic Anesthetic System
>
> 5. Investigations of Geopathogenic Factors and Their Effects on Inducing of Human Illness
>
> 6. Application of the Burr-Ravitz Electrodynamics Field Theory to the Precise Electronic Prediction and Determination of Female Ovulation Times in Terms of Minutes
>
> 7. Application of Newly Developed U. S. Army "Thermoviewer" Devices to Analyze Diagnostically the Status of Body Tissue Beneath the Skin

8. Investigations into the Use of Negative Ion Generators and Their Potential Application in Enhancing Human Performance, Welfare, and Health

9. Use of the Italian – Developed Device Referred to as the "Tobiscope" to Analyze and Detect Abnormal or Malignant Conditions in Body Cells and Tissues and to Locate Acupuncture Points

10. Use of High Frequency Electromagnetic Waves with Pulse Cadence Near Alpha Rhythm Brain Waves to Produce Narcosis Effects

11. Use of Psychobiological and Physico-Chemical Drugs to Alter States of Consciousness and Control Moods, Fatigue, Alertness, Personality, Perception, Tension

12. Through Interaction of Biophysiological Techniques and Electromagnetic Fields, Varying Emotional States of Groups of People and Inducing Visual or Auditory Hallucinations by Applying Specific Field Intensities

13. Luminescent Microscopy Techniques to Diagnose Illnesses

14. Analysis of Acupuncture and Its Degree of Effectiveness in Treating Human Ailments

15. Human Sensitivity to, and Effects of, Ultrasonic Energy

16. Biological Effects in Strong or Reinforced and Weak or Disturbed Geomagnetic and Electromagnetic Fields

17. Production of Electrographic Images of Living Organisms

18. Human Perception and Relation to the Conditioned Reflex

19. Applications of Neurocybernetics

20. Biofeedback Training and Applications

21. Electromagnetic Field and Magnetoresistance Effects on Neural Tissues

22. Biological Effects of Water Treated by Magnetic Fields on the Behavior and Activity of Living Organisms

23. Use of 'Microwave (SHE) Therapy' in the Treatment of Diseases

24. Investigation of the Stimulating Effect of Electromagnetic Fields on Hemopoiesis and on the Composition of the Blood in Humans and Animals

25. Measurement of the Electric Parameters of Body Tissues in Various Frequency. Ranges for Diagnostic Purposes

26. Effects of Electromagnetic Fields on Cultures of Normal and Malignant Human Cells and on Malignant Tumors

27. Effects of Electromagnetic Fields on Human Blood in Respect to Clotting Time and Erythrocyte Sedimentation Rate

28. Destruction of Bacteria by Electromagnetic Waves of Certain Wavelengths and by Biotherapy Means

29. Physical and Chemical Factors which Influence and Inhibit the Olfactory Senses

30. Ability of Electric Potentials on the Skin (referred to as "skin potentials") to Immediately Reflect Internal Changes in the Human Body

31. Use of Magnetic Fields and Magnets to Effect Pain-Deadening Conditions."[108]

I reviewed the resumes (CVs) of several of the people affiliated with the organization. It was a list of a highly creative and unusual group of professionals with diverse backgrounds allowing the productive "google-like" cross-fertilization that the company required. MRU also was known for working with individuals on the edge of sciences – people that were not taken seriously in many cases by mainstream science.

108. http://www.ajweberman.com/mankind2.htm This site contained a number of documents reported to have been removed from Mankind Research Unlimited's files. EPI6025

Many of the areas that they were concerned with were outside of the "accepted" disciplines. MRU recognized that good science innovation didn't usually spring from the status quo. In fact, it was through the connecting of scientists that the organization realized its potential and developed what appears to be some powerful research which could be used for the good of mankind – or its destruction. When I tried to locate any references to this company in terms of their current website none was found – the old sites were disabled. I did find that Dr. Schleicher, the President of Mankind Research Unlimited Incorporated, was operating for sometime after at the same address but under a different company name – the Foundation for Blood Irradiation.

Persinger and His Work

Inducing behavior rather than just reading a person's emotional state is the subject of one scientist's work in Canada. "Scientists are trying to recreate alien abductions in the laboratory... The experiment, to be run by Professor Michael Persinger, a neuroscientist at Laurentian University in Sudbury, Ontario, consists of a converted motorcycle helmet with solenoids on its sides that set up magnetic fields across a subject's head."[109] This experiment was carried out and was the subject of a Canadian Broadcasting Company (CBC-TV) exposé on mind control. The segment ran on a program called "Undercurrents" in February, 1999. I also appeared in that program, along with several others interested in this field of mind control.

For over 20 years Dr. Persinger, "has been working on a theory that connects not only UFOs and earthquakes, but also powerful electromagnetic fields and an explanation of paranormal beliefs in terms of unusual brain activity. He has

109. Watts, Susan. "Alien Kidnaps May Just be Mind Zaps." *Sydney Morning Herald*, Nov. 19, 1994. EPI816

also found that stimulating another area, the temporal lobes, can produce all sorts of mystical experiences, out-of-body sensations and other apparently paranormal phenomena."[110]

The work of this doctor suggests that these experiences may be the results of activity in the brain and not the actual experiences that are reported by individuals. He has had some measure of success in recreating many of these experiences in his subjects. Dr. Persinger is also known for his work in studying the effects of ELF on memory and brain function.[111]

In a paper[112] by Persinger, a great deal is revealed. What he says is extremely important. He indicates that the brain can be altered with very little power including that which is released from the natural geomagnetic activity of the earth or via contemporary communication networks. He suggests that within a very narrow set of variables we can stimulate the sense of smell, taste, touch, sight or hearing in a manner that would not permit us to see the difference between the laboratory created experience and reality. This is the direction of the science. What Dr. Persinger points out is that by using the earth's natural energy fields a signal could be generated at power levels that were consistent with the earth's and would be hidden in the "noise" created by the many manmade background radiating sources of energy.

On this grand scale, the use of mind control was contemplated as far back as 1969 by a former science advisor to President Johnson. "Gordon J.F. Macdonald, a geophysicist specializing in problems of warfare, has written that accurately

110. Opall, Barbara. "U. S. Explores Russian Mind-Control Technology." *Defense News*, Jan. 11-17, 1993. EPI818
111. Persinger, M. et al. "Partial Amnesia For a Narrative Following Application of Theta Frequency EM Fields." *Journal of Bioelectricity*, Vol. 4(2), pp. 481-494 (1985). EPI372
112. *On The Possibility of Directly Accessing Every Human Brain by Electromagnetic Induction of Fundamental Algorithms*, by M.A.Persinger, Laurentian University, June 1995, *Perceptual and Motor Skills*, June, 1995, 80,791-799, ISSN#0031-5125, EPI6026

timed, artificially excited strokes, 'could lead to a pattern of oscillations that produce relatively high power levels over certain regions of the earth...In this way, one could develop a system that would seriously impair the brain performance of very large populations in selected regions over an extended period.'"[113] This capability exists today through the use of systems which can stimulate the ionosphere to return a pulsed signal which at the right frequency (modulation) can override normal brain functions. By overriding the natural pulsations of the brain, chemical reactions are triggered, which alter the emotional state of targeted populations.

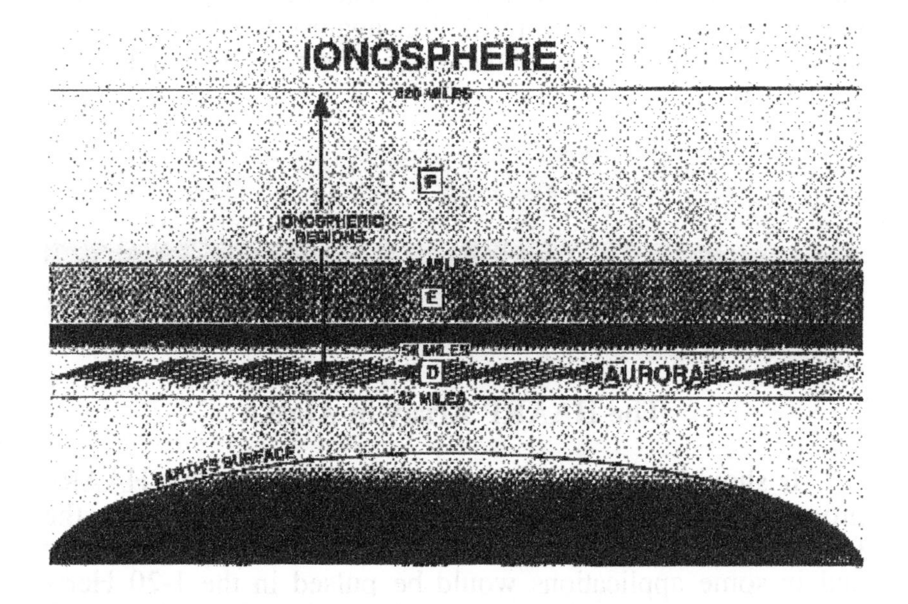

The fact that the United States government has a device that can cause the earth's magnetic fields to vibrate to any harmonic within a very large range through the use of

113. Brzezinski, Zbigniew. *Between Two Ages: America's Role in the Technetronic Era.* Viking Press, New York. 1970. EP1787

HAARP[114] has not been lost on many following these matters. The same could be accomplished through other signal carriers as well. The idea of controlling human beings is what the military is now, in part, referring to as "controlled effects".

The same issues being raised by the Russians in 2002, and pointed out by Dr. Persinger, were the subject of our own controversy in the United States in 1976, as a result of what was being referred to as the "woodpecker signal" being reported by short-wave radio operators. This signal originated out of the USSR long before the United States had developed this capability. The Russians knew what was possible in 2002, when they passed their resolution against HAARP, because they had created some of the same effects as far back as at least 1976.

The HAARP Array in Alaska

It should be kept in mind that John Heckscher, HAARP program manager, made clear in an interview that the frequencies and energies used in HAARP were controllable and in some applications would be pulsed in the 1-20 Hertz range. The ranges of frequency and at energy levels he stated, were small, but distinguishable from the Earth's own pulsations.[115] This point was further amplified by Heckscher

114. *Angels Don't Play This HAARP - Advances in Tesla Technology*, Begich & Manning, Earthpulse Press, 1995. EPI6027
115. Interview of John Heckscher , HAARP Program Manager, with Jeane Manning, 2-21-1995.

when he said, "The ELF/ULF waves to be generated by HAARP through interaction with the polar electrojet will have power levels so small compared with the existing background that special integrating receivers will have to be used to detect them."[116]

The idea of controlling the human mind was the focus of the work of Dr. Jose Delgado at Yale University from the 1960's through the 1980's. The issue we were concerned about regarding HAARP as they relate to his research is one of controlled coherent signals, which can be one fiftieth (1/50th) the energy level of the Earth's natural fields and still have profound effects on brain activity. The HAARP system creates a huge coherent controllable electromagnetic field which could be compared to a Delgado EMF (Electromagnetic Field), except it doesn't fill a room, it potentially fills a region the size of a large western state and, possibly, a hemisphere. Basically, the HAARP transmitter, when generating ELF pulse rates will emit energy of the same level as the Earth's, which is 50 times more than was needed in the wireless experiments of Dr. Delgado for affecting people. What this means is that if HAARP is tuned to the right pulse rate and modulated, using just the right wave forms, mental disruption throughout a region could occur intentionally or as a "side effect" of the transmissions.

Another important factor not thoroughly reviewed in the HAARP documents is cyclotron resonance. The HAARP signal uses this concept in its radiated energy. The effect of cyclotron resonance is a large increase in reactions in the ionosphere and in living organisms on the ground, under certain conditions. Many activities in living cells involve charged particles, and cyclotron resonance allows for the transfer of energy to cause

116. Letter to Mr. Arther Grey, Secretary to the United States Department of Commerce, National Telecommunications and Information Administration, Interdepartment Radio Advisory Committee, Spectrum Planning Committee, November 8, 1994.

ions to move more rapidly. It is cyclotron resonance which allows very low strength electromagnetic fields, together with earth's magnetic field, to produce significant biological effects. It occurs because the total effect has an impact on very specific particles when tuned to the right frequency codes. When combined with the Earth's normal magnetic fields, it is important to note, ELF frequencies (1-100 Hertz, pulses per second) appear to cause biological effects.

Cyclotron resonance can be visualized as a particle being spun like a coiled spring and then screwed right through the cell wall into the cell. This effect is one of the major considerations in electromedicine. To some degree this explains why non-ionizing levels of radiation produce the effects they do.[117] In other words, cyclotron resonance provides a condition where a significant interaction can take place, causing chemical reactions or other physiolog-ical responses. The manipulation of energy, when applied to living systems, can be used for enhancing the potentials of people or for harming them. It is here where the reactions occur leading to good, neutral or negative health effects.

The concept of cyclotron resonance was applied to the research carried out by the U. S. Naval Medical Research Center. The experimenters were able to apply external fields in such a way as to affect the brain chemistry of rats.[118] The same effects can be created in humans. The Navy research showed that they were able to affect the lithium ion, occurring naturally in the brain, so as to create the same effect as if they had treated the animal with a chemical introduction of lithium (lithium is used as a strong antidepressant).[119] Stated another way, you could say that by harmonizing or resonating with the frequency codes of naturally occurring chemicals, you could

117. *Cross Currents, The Perils of Electropollution, The Promise of Electromedicine*, by Robert O. Becker, M.D., pp 236-237.
118. Ibid.
119. Ibid, pp 236-240.

amplify their potency in the body of the animal, thereby creating the same chemical changes as would have occurred with a massive dose of the chemical being administered.

Other Considerations

The human body is driven by very subtle energy flows throughout its form. It is these underlying flows of energy that are measured as nerve impulses, signals running with blood flow, muscle movement, acupuncture meridians and points, and other energy signatures that can be detected and measured in the body and mind. Under every chemical reaction in physical form there is an energy exchange which can be measured and manipulated by external means. The physics of the energy interactions is the root to understanding and controlling the body and mind. The knowledge of these methods has advanced dramatically since the 1960's.

A brilliant electrophysiologist, Dr. Reijo Mäkelä, studied physics and chemistry in 1953-1959, followed by a two year study in psychology and additional studies in electrophysiology. He continued his education with advanced studies in Western and Chinese medicine and acupuncture, and practiced medicine as both a research scientist and physician for over forty years. Reijo was one of my mentors up until his death a few years ago.

Reijo developed a system of electro-laser acupuncture and trained a number of people in his methods prior to his death. His daughter, Dr. Anu Mäkelä,[120] now carries his work forward, while advancing her own outstanding and significant findings in these areas.

I spent a week with Reijo in his clinic in 1994. He had taken his electro-laser system into Finland and was challenged

120. Dr. Anu Mäkelä's research can be found at www.emred.fi

by their medical authorities, who are similar to our American Medical Association. They challenged him all the way to the highest court in Finland. And my friend, Dr. Mäkelä, won his case and was able to train fifteen practitioners in electro-laser methods before he passed away. That method is permitted by medical authorities in Finland and other European countries today, but used by very few practitioners within the United States.

Here's what Reijo did for me to demonstrate his technology. He first took a look at my general health utilizing the normal tools of an M.D. He checked my blood pressure, temperature and all the basic parameters used by most physicians. He analyzed my urine and blood, etc. Reijo was also an iridologist, someone that studies the iris and can connect disorders within the body to manifestations that appear in the iris. Iridology is well known. Anyone who studies this field will find that the body projects its state of health through the eyes. After Reijo analyzed, using both traditional western methods and nontraditional methods, he concluded that I had had pneumonia or some lung disorder at one time that caused me to have scarring in one lung, which was significant compared to the other lung. What he didn't know, was that several years before I had double pneumonia and was hospitalized. He had exactly diagnosed my condition. He explained that he would show me how electro-laser acupuncture works in manipulating energy in my body.

Using his electro-laser device Reijo first stimulated a lung point in my body. He wanted to show me how energy could come into the body and affect the primary lung points, which are on the chest. He said without coming anywhere near my chest area he could affect the lung points in my ear and it would show up on my chest. He was able to show energy coming through my body. Even with no physical contact, visible on my chest were two perfectly circular red dots about the size of a dime (centimeter). One was very red and the other one was a dull, pinkish color, correlating with which lung had the most damage and the lung which didn't. The energy coming in through the point in my ear started as a slight pulsing sensation at the ear point of entry. The feeling got more intense until it was a rhythmic, pinching sensation. Then it became so intense I couldn't stand it anymore and I had him take the instrument away from my ear.

In looking at my general physiology, he could tell my liver was in excellent shape so he went through a liver point, never adjusting the settings on the laser instrument as far as intensity. On that same setting going through the liver, which was functioning normally, there was no sense of pain, just that slight oscillating, pulsing sensation, but it never elevated. It never increased to be uncomfortable. The reason it never increased is because the energy flowed through the acupuncture meridians as it was intended, not getting blocked. Where there is resistance, the blockage actually sent the energy back to the point of entry, creating an irritation on the nerve endings at that point of entry. That was what caused the pain in my ear points when he was trying to affect my lungs. It was interesting to see the physical manifestation, the energy delivered where those primary lung points clearly showed it.

How does this process work in treating patients? Before Reijo died he treated over 16,000 patients with a number of different disorders. Patients would go in every day for a week to eight weeks of treatments depending on the disorder. These lasted 30 minutes and they got the same basic treatment five days a week. This system of acupuncture generally does not work instantly. In many cases it takes time. Many things that Reijo was treating are considered incurable in the west. He successfully treated MS, diabetes, and various forms of cancer. (When people say they've got a cure for cancer, be a little bit skeptical in the sense that there are over 700 types of cancer.

People need to be a little bit more specific.) Reijo had success with many specific disorders and forms of cancer within the body.

Reijo's work is being carried forward by his daughter and others but it didn't just include electro-laser acupuncture. The work also included the proper nutrient supplements and the proper foods in order to make sure that the building blocks for the body were also present as the patient rebalances his system so that energy flows correctly.

During my 1994 visit with Dr. Mäkelä, I discussed the HAARP project. The idea that the HAARP system could be used as a transmitting system for brain entrainment was explored, and it reminded Dr. Mäkelä of a number of FFR devices. The idea that HAARP could create various waveforms across a large segment of the electromagnetic spectrum was of concern, considering the health risks. This system had the possibility of being misused, and the risks had not been well defined. Inadvertently there could be negative health effects if the operators were unaware of the consequences of pulsed electromagnetic radiations in the wave forms they were using – energy waves which were being bounced back to the earth from the ionosphere.

In particular, when reviewing the HAARP materials, Dr. Mäkelä was reminded of a device which was marketed for a short time in Japan. This device was a radio transmitter which carried a pulsed wave (6-12 Hertz). The pulses were in resonance with the radio waves, so that the radio waves acted as a carrier wave to create the desired effect. The device was used to relax or focus attention. The effect was created wirelessly and at a distance.[121] It should be noted that a United States military representative has already acknowledged that

121. Dr. Nick Begich visits and conversations with Dr. Reijo Mäkelä in Finland, November, 1994.

they intend to use HAARP to transmit pulsed radio frequency radiation (which we assert, if modulated in just the right way and with the right waveform, could cause serious negative health effects on people and animals). HAARP specifications call for a very versatile set of parameters[122] which can do what has otherwise only been demonstrated in the laboratory or under controlled conditions – disrupt minds and create negative health effects.

The United States military is very familiar with these technologies. Captain Paul Tyler was the director of the U. S. Navy's Electromagnetic Radiation Project from 1970 through 1977. He was quoted in a February, 1985, *Omni* article about the effects of electromagnetic radiations. He concluded that effects which could be stimulated chemically could also be stimulated electrically. "With the right electromagnetic field, for example, you might be able to produce the same effects as psychoactive drugs."[123] The ideas first kicked around by the CIA and early researchers were now being pursued by the military for use in controlling human behavior – a prospect with profound implications.

Another player in this seemingly disconnected group was Arthur Guy. Under contract to the United States Air Force, he helped compile the Radiofrequency Radiation Dosimetry Handbook. The book is the only one we could find in the open literature that explores a number of critical areas needed for further developments in the use of radio frequency energy to affect people. In that publication, Guy's work from the University of Washington was described. In this study he exposed rats to low levels of electromagnetic radiation. The effect of these exposures included immunological stress and increases in the formation of tumors – four times the rate of

122. Office of Naval Research, Contract Number N00014-92-C-0210, ARCO Power Technologies Incorporated as Contractors, September 16, 1992, with Amendments through October 19, 1993.
123. "The Mind Fields", by Kathleen McAuliffe, Omni Magazine, February 1985.

unexposed animals. This study produced negative effects at radiation levels twenty times below the established United States safety thermal level![124]

Additional research in non-thermal effects is being conducted throughout the world. In Germany, the Deutsche Forschungsgemeinschaft – the equivalent of the American Academy of Sciences – announced their research results concerning ELF. They concluded that "nonthermal effects due to EMF exposures can be triggered in living cells under selected conditions." Reportedly, their research in these areas would continue.[125]

The Woodpecker Signal

In the late 1990's I obtained a copy of what appears to be an unpublished manuscript by research scientist Dr. Andrija Puharich[126]. Dr. Puharich had a long history of both private and government research projects and was one of the creative scientists who was involved in the early days of this research. The document revealed that the use of radio signals being sent out by the Soviets on July 6, 1976 was a major problem. The USSR had filled the entire planet with "radio noise" from transmitters based in their country. Canadian officials gave Puharich the data on these ELF signals for his analysis. Puharich looked at the signal being broadcast and determined that it had a frequency of 5-15 pulses per second (Hertz). The signal could create a steerable beam 42 miles wide that could sweep a great circle route around the planet passing selectively, for example, through Ottawa, New York, Washington, D.C. and so on. The amount of energy involved was very small (25 nanoteslas). This is a very, very small

124. *Cross Currents, The Perils of Electropollution, The Promise of Electromedicine*, by Robert O. Becker, M.D., pp 194-197.
125. German Workshop on Mechanisms of EMF Interactions, *Microwave News*, November/December, 1991.
126. *Time No Longer*, by Andrija Puharich, 1979, pp 242-251. EPI6048

amount of energy lost in the natural background noise if it were not for its controlled signal.

From the literature, Puharich knew that it was the magnetic component of the electromagnetic field that was psychoactive and could cause mood or behavioral changes. Puharich, together with Robert C. Beck, designed and built devices to see what would happen if the signals were created in the lab at 10 to 100 nanoteslas and then connected an electroencephalograph (EEG) to measure brain waves which were then displayed on a dual channel oscilloscope. The other oscilloscope measured the artificial ELF signal so that observations could be made in the 2-20 Hz. range (pulses per second). What they observed in a controlled test was that 30% of the test subjects showed brain wave entrainment by the ELF signals with 50% showing pyschophysiological reactions characteristic of specific ELF frequencies. Entrainment is when the brain mirrors the pulse rate of the external artificial signal thereby causing changes in body and mind. It was found that people are ultra-sensitive to very weak magnetic fields. Brain entrainment is the same as the Frequency Following Response (FFR) mentioned earlier. Several effects were reported because of these weak signal transmissions including the following:

– 6 Hz. – Headaches were reported.

– 6.66 Hz. and lower – Nausea, headaches, confusion, and depressive anxiety.

– 7.8 Hz., 8 Hz., 9 Hz. – Produced alpha rhythms often reached in meditation and a sense of well-being.

– 10.35 Hz. – Created agitated anxiety, fear and hostile aggressive behavior.

– 11 Hz. – Induced riotous behavior.

Puharich and Beck made one other finding – that there was no protection from a system like this because the signals themselves passed through everything on the earth. Again we draw readers back to the issue of HAARP because this is the biggest, most flexible system for creating these kinds of global effects and is far superior to what the Soviets had in 1976.

LIDA: The Soviet Device

In the early part of the Vietnam War, in the mid-1960's, a device called the LIDA machine was captured from the Russians. It was used to interrogate U. S. prisoners in Vietnam. Essentially it was an oscillating electromagnetic signal working with a strobe light mixed with auditory signals that caused brain entrainment. It was used for putting people in a trancelike state for interrogations. This put people in a highly suggestive state where they were more willing to give information. The device was also reported to incorporate "40 megahertz radio waves which stimulate the brain's electromagnetic activity at substantially lower frequencies."[127] The author of this article was affiliated with Mankind Research Unlimited (MRU) and has extensive other background that gives special credibility to this article.[128] The device was a

127. Psy-war: Soviet Device Experiment, by Dr. Stephan Possony, Associate Editor, *Defense & Foreign Affairs Daily*, Vol. XII, No.104, June 7, 1983. EPI6046

128. STEFAN T. POSSONY – Sovietologist and Psychological War-fare Specialist Dr. Possony is a Sovietologist, International Affairs, and Psychological Warfare Specialist of high reputation and experience. Prior to and during the early stages of World War II, he served as an Advisor to the French Air Ministry and the French Foreign Office. After this assignment, he came to the United States and held a post as a Carnegie Research Fellow at the Institute for Advanced Study, Princeton, New Jersey. During World War II and through 1946, he was a Psychological Warfare Specialist at the Office of Naval Intelligence (ONI). Between 1946 and 1961, he served as Special Advisor to the Assistant Chief of Staff, Intelligence, USAF. During the same period, Dr. Possony served as Professor of International Politics, Georgetown University, and during 1956-1958 as Director of Research for Life Magazine's Russian Revolution project. In 1961 Dr. Possony became Director of the International Political Studies Program at the Hoover Institution on War, Revolution and Peace, where he is now a Senior Fellow. He testified before the U. S. Senate Internal Security Sub- Committee on the "Threat of U. S. Security Posed by Stepped-up Sino-Soviet Hostilities" and has on frequent occasion been called upon as a special consultant to U. S. Presidential Committees, Congress, and the Defense Department.

powerful technology that added to the advances that were being made in the United States.

Adey was the only U. S. scientist to test this Russian machine at that time. He was one of the government's and industry's often cited experts on the effects of electromagnetic fields on people and other living things. Combined with other stimuli, the LIDA machine has also been reported to induce deep sleep. When interviewed, Adey acknowledged that the device worked, but refused to comment on the United States military's use of this technology.[129 130] Dr. Adey traveled to the Soviet Union in the early 1970's giving him an opportunity to look into these concepts in greater depth. What Adey's research indicated in 1974 was significant. "New and exciting findings indicate a clear susceptibility of the mammalian brain to extra-low-frequency (ELF) fields at brain wave frequencies, with extremely weak gradients in the air. Similar, more powerful interactions, have been found when VHF radio fields are amplitude-modulated at EEG frequencies, including effects on brain calcium binding. The findings suggest the use of nonionizing electromagnetic fields, with a variety of modulation patterns that stimulate the frequencies of brain neuroelectric activity, as powerful new tools in brain research."[131]

While a neuroscientist at the Loma Linda, California, Veterans Administration Hospital, Adey demonstrated that the brain waves of animals can be manipulated. He showed that they can be locked in phase with external pulsating waves, which then affect subtle changes in behavior including enhanced learning. He further demonstrated, in work with Dr.

129. The Mind Fields", by Kathleen McAuliffe, Omni Magazine, February, 1985.
130. Resonance, Newsletter of the Bioelectromagnetics SIG, Number 28, May, 1995, Judy Wall, Editor, 684 C.R. 535, Sumterville, Florida, USA. (This is a special Interest Group of M.E.N.S.A.)
131. Brain Research Institute, University of California Los Angeles 13th Annual Report July 1,1973 to June 30, 1974 Page 72. EPI6047

Suzanne Bawin, that fields too weak to trigger a nerve impulse can alter the way calcium ions bind to cell surfaces, causing an array of chemical reactions within the cell itself.[132] What this means is that the kind of chemical changes which are needed to cause alterations in thinking or health can be started by very small amounts of controlled power, through understood and manipulatable means.

The LIDA device and other devices triggered a whole cascade of interest, primarily from the Central Intelligence Agency. It was right after this period, in the early 1970's, that the adventure became of interest to the United States Congress. In 1975, the Congress authorized the Rockefeller Commission to conduct a full investigation of what the Central Intelligence Agency was doing in the United States. At the conclusion of the investigation they published a report on the activities of the CIA within the United States.[133] The Central Intelligence Agency was supposed to only be dealing with things outside of the United States, not within its boundaries. In the 1960's the CIA, according to the report, was infiltrating student and other groups objecting to U. S. policies at the time. They were reported to be involved in surveillance of everyone from the Congress all the way through activist organizations and individuals throughout the United States, and they were using unwitting test subjects for mind control experiments.

The report to the President disclosed that there had been thousands of victims. The sad story for all of the victims, including thousands of military personnel used in these experiments, was that the government was never fully held accountable until one person actually committed suicide. That resulted in a lawsuit that eventually was settled by the Central

132. "The Mind Fields", by Kathleen McAuliffe, Omni Magazine, February, 1985.
133. Report to the President by the Commission on CIA Activities Within the United States, June, 1975. EPI6014

Intelligence Agency in favor of that individual, someone who had been subjected to LSD experiments. There were surely more victims whom we will never hear about until all of the files are opened and made public.

Unfortunately the reverse is now happening, and increasing amounts of material are being classified and hidden by governments. In 2004, the United States government spent $7.2 billion to classify 15.6 million documents, which is almost double the 8.6 million records so classified in 2001. On the other end of the spectrum, in 2004, there were 28.4 million documents declassified down from 100 million in 2001, and the 204 million documents declassified in 1997.[134] In 2006 even more significant information is held and never released while piles of files are classified, further limiting government accountability.

In a cascade of controversy at the end of 2005 and in 2006, it was reported that President Bush had authorized illegal wire taps,[135] even after passage and use of the Patriot Act. That Act of Congress went so far over the edge that it was objected to by hundreds of local governments around the United States. But it was not enough for the "intelligence community" in the United States. The President was reported to have admitted authorizing the National Security Agency (NSA) to conduct eavesdropping without court orders, using fear of terrorism as the excuse. This secret Executive Order was authorized and reviewed every 45 days, with Bush reauthorizing this illegal activity thirty times.[136] Where has the Congress been in all of this besides hiding behind poorly

134. "Report: Federal Secrecy Expands, Grows Expensive", by Michael J. Sniffen, The Associated Press, printed in the *Anchorage Daily News*, September 4, 2005, pp A-3. EPR6055

135. "Stateside Surveillance Ordered by President Fans Firestorm, Members of Congress Decry Secret Plan Signed by Bush", by Dan Eggen & Charles Lane, *The Washington Post*, reprinted *Anchorage Daily News* December 17, 2005. EPI6056

136. "Bush Admits he Ordered Domestic Spying Program", by David Sanger, *The New York Times*, reprinted *Anchorage Daily News*, December 18, 2005. EPI6057

executed power? This takes us back to Nixon and his administration. By the middle of 2006, the cards were beginning to fall with more to come for sure. All of the issues from surveillance by the NSA, to the CIA's transporting people[137] to the torture chambers of the third-world, and the indictment of some in Congress with more predicted by the mass media to come. Within days both the Democrats and Republicans in the Congress were calling for an investigation into the activities of the Bush White House.[138]

Another interesting historical note is that the Foreign Intelligence Surveillance Act (FISA) which was established in 1978, after the Presidential Report on the abuses of the CIA, and because of President Nixon's use of the CIA in a number of clearly illegal activities. As a part of this Act, a special court was set up that has operated since that time and stood up to the U. S. Constitution. In over 28 years the court has only denied a warrant five times. This has been good enough for both Republican and Democratic Presidents. It is this Act that precludes what President Bush is doing in 2006,[139] which is clearly illegal under U. S. law and not yielding much in the way of results. To date FBI agents have been pulled off of legitimate criminal work to chase the "leads" of the NSA, which after thousands of investigations have yielded virtually nothing. At the same time the implications of each public revelation of abuses of Presidential power get progressively clearer.

137. "White House's Surprises Raise Oversight Doubt", by Dana Milbank, *The Washington Post*, reprinted in the *Anchorage Daily News*, December 18, 2005. EPI6058
138. "Lawmakers Decry Spying Program-NSA:Democrats, Republicans call for Probe into President's Authorization of Eavesdropping", by Hope Yen, The Associated Press, reprinted *Anchorage Daily News*, December 19, 2005. EPI6059
139. "Why NSA spying Puts the US in Danger, A Former Analyst Look's at the Agency's Current Controversy", by Ira Winkler, *Computerworld*, May 16, 2006. EPI6063

Dr. Jose Delgado - Mapping the Brain

Dr. Jose M. R. Delgado, M.D., has researched the human brain and published his results in professional papers since 1952. He became a Professor of Physiology at Yale, where he continued his research work focusing on mapping and controlling the brain. His early work pioneered our understanding of the human brain and was summarized through 1969 in a book he wrote, called *Physical Control of the Mind; Toward a Psychocivilized Society.*[140] This early work was mainly animal research, and included insertion of electrodes into the brains of animals. He found that by inducing an electrical current in the brain of his subject, he could manipulate behavior. Delgado discovered that he could create a number of effects from sleep to highly agitated states of consciousness. His later work was done wirelessly, with his most advanced efforts developed without electrode implants used at all. Stated differently, he achieved the brain manipulating effects at a distance, without any physical contact or devices attached to the living creature being manipulated. Dr. Delgado found that by changing the frequency, pulse rate and waveform on an experimental subject, he could completely change their thinking and emotional state. Dr. Delgado's work set the foundation for many other researchers, while at the same time opening the possibilities of misuse by the government. In Dr. Delgado's book, he uses a quote from the UNESCO's Constitution which, in the context of his work and this book, is foreboding: "Since war begins in the mind of men, it is in the minds of men that the defenses of peace must be constructed."[141] It is a thought, in the context of Delgado's work, which smacks of the Orwellian story, *1984.*

140. *Physical Control of the Mind; Toward a Psychocivilized Society*, by Jose M.R. Delgado, M.D., 1969.
141. United Nations Educational, Scientific and Cultural Organization's Constitution.

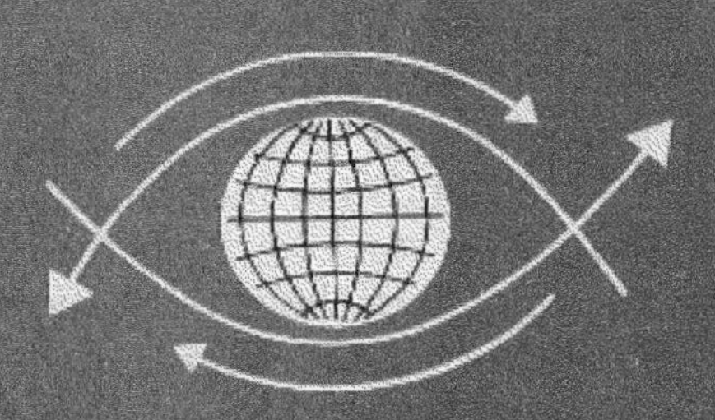

Dr. Jose Delgado's Book

Delgado was able to stimulate the brain utilizing an implant technology, which caused a number of effects in primates, bulls and even human beings. By the 1980's he found that he didn't need any physical contact with the human brain. He just needed to oscillate, or vibrate, energy into the brain in a very specific way. He discovered in 1985 that you could create tremendous changes in human brain chemistry by oscillating energy at 1/50 of what the Earth naturally produces in the radio frequency range.

Actual testing of certain systems proved "that movements, sensations, emotions, desires, ideas, and a variety of psychological phenomena may be induced, inhibited, or modified by electrical stimulation of specific areas of the brain."[142] By 1985, Dr. Delgado was able to create these effects using only a radio signal sent to the brain remotely, using energy concentrations of less than 1/50th of what the Earth naturally produces. This discovery implied that frequency, waveform and pulse rate were the important factors, rather than the amount of energy being used. In considering this, it makes sense because the human body does not require high electromagnetic power concentrations to regulate its normal functioning – the key was in finding the "tuning" mechanisms for locating the right "receiving station" in the brain and body of a test subject.

If we consider that, in a broader sense we might ask, "how much radio frequency energy is around us right now?" At this moment there is approximately 200 million times more than nature produces. If you take what nature produces on its own, cut it down to 1/50th of that amount of energy, or one ten-billionth of the energy man creates, is sufficient to override human brains in such a way as to change them from lethargic and passive to highly aggressive and agitated – it was almost like throwing a light switch on and off, according to Delgado.

142. Delgado, Jose M.R. *Physical Control of the Mind: Toward a Psychocivilized Society.* Harper & Row, Publishers. New York, 1969. EPI850

In a book about the Central Intelligence Agency's (CIA) pursuit of mind control technology, Dr. Delgado's work is also discussed.[143] Dr. Delgado's research was reviewed in 1969 by Dr. Gottlieb, who worked for the CIA's Office of Research and Development (ORD) while looking into the possible uses of this technology. At that time, the work was still crude, although the CIA shared Dr. Delgado's vision for techniques which would allow for a "psychocivilized society".

During this period, a neurosurgeon at Tulane University, Dr. Robert G. Heath, brought this prospect closer to reality with his work in electrical stimulation of the brain. As one author recently put it, "Like Dr. Delgado, the neurosurgeon concluded that ECS (Electro-Cranial Stimulation of the Brain) could evoke hallucinations as well as fear and pleasure. It could literally manipulate the human will at will."[144]

To the CIA, Dr. Delgado's "wireless" effects were thought-provoking. He discovered that animals could be placed within an electromagnetic field (EMF) and be manipulated without any physical contact. "The fields Delgado uses are as low as one fiftieth the strength of the Earth's own magnetic fields...yet when the signal is tuned to precise frequencies, Delgado can do much more than make a monkey sleepy."[145] These technologies are increasingly being recognized by other researchers, and the database of available frequencies and inducible behaviors, or other changes, has reached a level of predictability not possible even a few years ago.

143. *Journey into Madness, The True Story of Secret CIA Mind Control and Medical Abuse*, by Gordon Thomas, 1989, pp 276-279.
144. Ibid., pp 276.
145. "The Mind Fields", by Kathleen McAuliffe, *Omni Magazine*, February 1985.

Chapter Four

Mind Wars

A lengthy 1993 article in the Wall Street Journal discussed the direction of the military in the development of what they were calling at the time, "Nonlethal" weapons. The article explored the use of a new class of weapons which could be used to disrupt communications, radars and other electronic equipment. In the article, retired Lt. General Richard Trefry, a military advisor to President Bush, said, "They're all real." He then went on to say, "But you're bordering on classified stuff here."[146] The article described a series of new developments leading to these technologies which did not exist ten years before. Time schedules for the development of these technologies had been established in 1982, with the systems expected to be ready for use in the early to middle 1990's.[147]

"For the first time in some 500 years, a scientific revolution has begun that will fundamentally change the world as much as the Renaissance and Enlightenment did. A handful of extraordinary new advances in science are taking humans

146. "Nonlethal Arms, New Class of Weapons Could Incapacitate Foe Yet Limit Casualties", *The Wall Street Journal*, by Thomas E. Ricks, January 4, 1993, pp A1 and A4.
147. *Final Report On Biotechnology Research Requirements For Aeronautical Systems Through the Year 2000, Volumes I and II*, Southwest Research Institute, San Antonio, Texas, 1982.

quickly and deeply into areas that will have profound implications for the future."[148] One of these areas is control of the human mind. The issues surrounding behavior modification, mind control and information warfare become crystal clear as the facts unfold. The following is taken from a U. S. military document which clarifies their position in the emerging area of research, taking a direction that moves the technology to only military uses and possible abuses:

The Information Revolution and
The Future Air Force
Colonel John A. Warden, III, USAF

"We're currently experiencing, on an unprecedented global basis, three simultaneous revolutions, any one of which would be more than enough to shock and confound us. The first revolution, a geopolitical revolution, sees a single dominant power in the world for the first time since the fall of Rome. The opportunities that are inherent in this situation are extraordinary, as are the pitfalls. Unfortunately, there is no one around that has first hand experience in how to deal with that kind of single power dominant world.

The second revolution, and there's a lot of discussion about this so far, is the information revolution. As other people have mentioned, it is following inexorably in tandem behind Moore's law of computing power. Attendant to it, though, is not the creation of new ideas and technologies, but also an exponential growth in the velocity of information dissemination, and for us, that is of extraordinary importance. A key part of this information revolution has an awesome impact on competition. The business that introduced a new product ten years ago

148 Petersen, John L. *The Road To 2015: Profiles of the Future.* Waite Group Press, 1994. ISBN 1-878739-85-9. EPI849

could count on probably five years before it had to look seriously at potential competitors based overseas. Today, you're lucky if you can count on five months or even five weeks before you are facing the overseas competitor. In today's world, success simply demands rapid introduction of successively new products or military systems. Success now goes to the organization which exploits information almost instantly, while failure is the near certain fate of the organization which tries to husband or hide ideas. Real simple – use it or you're going to lose it.

The third revolution, which is a little bit more complex, is the military/technological revolution, or in some places called the revolution in military affairs. I'm convinced that this is the first military technological revolution ever because we now have, for the first time, a conceptually different way to wage war. We can wage war in parallel now. In the past, communications and weapons technology, especially weapons accuracy, have constrained us to waging serial war. This changes almost everything.

Biological Process Control: As we look forward to the future, it seems likely that this nation will be involved in multiple conflicts where our military forces increasingly will be placed in situations where the application of full force capabilities of our military cannot be applied. We will be involved intimately with hostile populations in situations where the application of Nonlethal force will be the tactical or political preference. It appears likely that there are a number of physical agents that might actively, but largely benignly, interact or interfere with biological processes in an adversary in a manner that will provide our armed forces the tools to control these adversaries without extensive loss of life or property. These physical agents could include acoustic fields, optical fields, electromagnetic fields, and combinations thereof. This paper will address only the prospect of physical regulation of biological processes using electromagnetic fields.

Prior to the mid-21st century, there will be a virtual explosion of knowledge in the field of neuroscience. We will have achieved a clear understanding of how the human brain works, how it really controls the various functions of the body, and how it can be manipulated (both positively and negatively). One can envision the development of electromagnetic energy sources, the output of which can be pulsed, shaped, and focused, that can couple with the human body in a fashion that will allow one to prevent voluntary muscular movements, control emotions (and thus actions), produce sleep, transmit suggestions, interfere with both short-term and long-term memory, produce an experience set, and delete an experience set. This will open the door for the development of some novel capabilities that can be used in armed conflict, in terrorist/hostage situations, and in training. New weapons that offer the opportunity of control of an adversary without resorting to a lethal situation or to collateral casualties can be developed around this concept. This would offer significant improvements in the capabilities of our special operation forces. Initial experimentation should be focused on the interaction of electromagnetic energy and the neuromuscular junctions involved in voluntary muscle control. Theories need to be developed, modeled, and tested in experimental preparations. Early testing using in vitro cell cultures of neural networks could provide the focus for more definitive intact animal testing. If successful, one could envision a weapon that would render an opponent incapable of taking any meaningful action involving any higher motor skills, (e.g. using weapons, operating tracking systems). The prospect of a weapon to accomplish this when targeted against an individual target is reasonable; the prospect of a weapon effective against a massed force would seem to be more remote. Use of such a device in an enclosed area against multiple targets (hostage situation) may be more difficult than an individual target system, but probably feasible.

It would also appear to be possible to create high fidelity speech in the human body, raising the possibility of covert suggestion and psychological direction. When a high power microwave pulse in the gigahertz range strikes the human body, a very small temperature perturbation occurs. This is associated with a sudden expansion of the slightly heated tissue. This expansion is fast enough to produce an acoustic wave. If a pulse stream is used, it should be possible to create an internal acoustic field in the 5-15 kilohertz range, which is audible. Thus, it may be possible to 'talk' to selected adversaries in a fashion that would be most disturbing to them.

In comparison to the discussion in the paragraphs above, the concept of imprinting an experience set is highly speculative, but nonetheless, highly exciting. Modern electromagnetic scattering theory raises the prospect that ultra short pulse scattering through the human brain can result in reflected signals that can be used to construct a reliable estimate of the degree of central nervous system arousal. The concept behind this 'remote EEG' is to scatter off of action potentials or ensembles of action potentials in major central nervous system tracts. Assuming we will understand how our skills are imprinted and recalled, it might be possible to take this concept one step further and duplicate the experience set in another individual. The prospect of providing a 'been there – done that' knowledge base could provide a revolutionary change in our approach to specialized training. How this can be done or even if it can be done are significant unknowns. The impact of success would boggle the mind!"[149]

The above report was a forecast for the year 2020. However the reality is that most of the enabling technologies already exist, and there are a number of patents in the open literature which clearly show the possibilities. This research is not new but goes back to the 1950's. "A new class of weapons, based on electromagnetic fields, has been added to the muscles of the military organism. The C^3I (Command, Control, Communications and Information) doctrine is still growing and expanding. It would appear that the military may yet be able to completely control the minds of the civilian population."[150]The targeting of civilian populations by the military is a significant departure from its history. In the past the military has used persuasion through real information rather than using deliberate deception and mind manipulation to win populations over. "A decoy and deception concept presently being considered is to remotely create the perception of noise in the heads of personnel by exposing them to low power, pulsed microwaves. When people are illuminated with properly modulated low power microwaves the sensation is reported as a buzzing, clicking, or hissing which seems to originate (regardless of the person's position in the field) within or just behind the head. The phenomena occurs at average power densities as low as a few microwatts per square centimeter with carrier frequencies from .3 to 3.0 GHz. By proper choice of pulse characteristics, intelligible speech may be created. Before this technique may be extended and used for military applications, an understanding of the basic principles must be developed. Such an understanding is not only required to optimize the use of the concept for camouflage, decoy, and deception operations, but is required to properly assess safety

149. *New World Vistas: Air And Space Power For The 21st Century – Ancillary Volume*, USAF Scientific Advisory Board. 1996. EPI402
150. *Summary and Results of the April 26-27, 1993 Radiofrequency Radiation Conference, Volume 2: Papers*. 402-R-95-011, U.S.EPA, March 1995. EPI728

factors of such microwave exposure."[151] Actual testing of certain systems has proven "that movements, sensations, emotions, desires, ideas, and a variety of psychological phenomena may be induced, inhibited, or modified by electrical stimulation of specific areas of the brain. These facts have changed the classical philosophical concept that the mind was beyond experimental reach."[152]

The ethical considerations have not changed, but the military's position on the ethics has changed as they have gained significant capabilities in these areas. "Psychological warfare is becoming increasingly important for U. S. forces as they engage in peacekeeping operations. 'In the psychological operations area, we're always looking to build on our existing technologies, so much of this is evolutionary,' Holmes said. 'It is critically important that we stay ahead of the technology curve.'"[153] The temptation to dabble in this area has once again overcome the ethical considerations.

The thought of these technologies being covertly used in the advent of a new concept in warfare was introduced by G. W. Bush after the 9/11/2001 incident, when he coined the phase of "preemptive warfare". This concept can be turned on us, the "civilian population", through the use of the kinds of technologies suggested in this book. There have been several incidences referenced where these technologies have been used. The most important aspect of the technology of energy-based systems is that they are at the speed of light – *nearly immediate*. As each of these technologies evolve, the concept of "preemptive warfare" needs to be reassessed in all of its implications for the world. In whose ideology, whose

151 Oscar, K.J. *Effects of Low Power Microwaves on the Local Cerebral Bood Flow of Conscious Rats.* Army Mobility Equipment Command. June 1, 1980. EPI1195
152. Delgado, Jose M.R. *Physical Control of the Mind: Toward a Psychocivilized Society.* Harper & Row, Publishers, New York, 1969. EPI850
153. Cooper, Pat. "U. S. Enhances Mind Games." *Defense News,* April 17-23, 1995. EPI1154

religion, who's economic system, whose reality will the world live, when some government, corporation or individual chooses to launch a preemptive war on the world or just some part of it? Which country or person will master the technology and who's ideology will be pressed into the consciousness of the people on the planet by its misuse? Will the technology lead to a more developed human that can manifest his existence in the fullness of his creative capacity, as a soul created in the image of the Creator[154] or as a slave to new misused technologies?

These weapons aim to control or alter the psyche, or to attack the various sensory and data-processing systems of the human organism. In both cases, the goal is to confuse or destroy the signals that normally keep the body in equilibrium.

According to a Department of Defense Directive, information warfare is defined as "an information operation conducted during time of crisis or conflict to achieve or promote specific objectives over a specific adversary or adversaries." An information operation is defined in the same directive as "actions taken to affect adversary information and information systems." These "information systems" lie at the heart of the modernization effort of the U. S. armed forces and manifest themselves as hardware, software, communications capabilities, and highly trained people.

"Information warfare has tended to ignore the role of the human body as an information or data processor, in this quest for dominance, except in those cases where an individual's logic or rational thought may be upset via disinformation or deception. Yet, the body is capable not only of being deceived, manipulated, or misinformed but also shut down or destroyed – just as any other data-processing system. The "data" the body receives from external sources – such as electromagnetic,

154. Genisis 1:27, Bible.

vortex, or acoustic energy waves – or creates through its own electrical or chemical stimuli, can be manipulated or changed just as the data (information) in any hardware system can be altered. If the ultimate target of information warfare is the information-dependent process, whether human or automated, then the definition implies that human data-processing of internal and external signals can clearly be considered an aspect of information warfare."[155]

155 "The Mind Has No Firewall." *Parameters*, Vol. XXVIII, No. 1, Thomas, Timothy L., Spring, 1998. EPI525

Chapter Five

Auditory Effects

The questions which this section raises are profound. Is it possible to transmit a signal to the brain of a person, from a distance, which deposits specific sounds, voice or other information which is understandable to them? Is it possible to transfer sound in a way where only the targeted person can hear the "voice in the head" and no one else hears a thing? Is it possible to shift a person's emotions using remote electromagnetic tools? The answer to each of these questions is a resounding – Yes! The state of the science has passed even the most optimistic predictions, and the capabilities are here now.

Even military literature suggests that this is possible. A series of experiments, patents and independent research confirms that this technology exists today. While giving testimony to the European Parliament in 1998[156] , I personally demonstrated one such device to the astonishment of those in attendance. This particular device required physical contact in

156. "HAARP and Nonlethal Weapons", February 5, 1998, hearing in the Committee on Foreign Affairs, Security and Defence Policy: Subcommittee on Security and Disarmament, European Parliament, Brussels, Belgium, Testimony given by Begich, Bertell, Lutz, and others.

order to work and was nearly forty years old in design. This area is one of the most important because it points to the ultimate weapons of political control – the ability to place information directly into the human brain, bypassing all normal filtering mechanisms. Electronic telepathy.

The Department of Defense put forward the following contract award in 1995, which would be used for direct communications with military personnel. "Communicating Via the Microwave Auditory Effect; Awarding Agency: Department of Defense; SBIR Contract Number: F41624-95-C-9007." The description of this technology was written as follows:

> **"Title: Communicating Via the Microwave Auditory Effect**
>
> Description: An innovative and revolutionary technology is described that offers a means of low-probability-of-intercept Radio Frequency (RF) communications. The feasibility of the concept has been established using both a low intensity laboratory system and a high power RF transmitter. Numerous military applications exist in areas of search and rescue, security and special operations."[157]

The feasibility was not only demonstrated in the laboratory but also in the field using a radio frequency carrier. In the case of the Gulf War, I had always suspected that the reason the Iraqis gave up en mass was not because of the heavy bombardments but because they were being hit with new "Nonlethal" systems that created fear and perhaps even worse. My research uncovered reports which now confirm my suspicions.

157. Dept. of Defense (awarding Agency). "Communicating Via the Microwave Auditory Effect." SBIR Contract Number: F41624-95-C-9007. EPI277

"What the 'Voice of the Gulf' began broadcasting, along with prayers from the Koran and testimonials from well-treated Iraqi prisoners, was precise information on the units to be bombed each day, along with a new, silent psychological technique which induced thoughts of great fear in each soldier's mind..."[158] This makes a great deal of sense today given what has become increasingly known about mind control weapons. "According to statements made by captured and deserting Iraqi soldiers, however, the most devastating and demoralizing programming was the first known military use of the new, high tech type of subliminal messages referred to as ultra-high-frequency 'Silent Sounds' or 'Silent Subliminals.'"[159] The use of these new techniques went well beyond the injection of fear and may have involved more powerful signal generators, which caused the other symptoms that the world observed including head pain, bleeding from the nose, disorientation and nausea – all possible with so-called nonlethal weapons. The project "Commando Solo"[160], run by the United States Air Force, is a C-130 aircraft that is used as a broadcasting platform sent into war zones in advance of troops to condition the adversary with propaganda of the traditional type and also to deliver other mind/emotion altering technology literally carried on the airwaves. These questions now remain: Are they still using the techniques like an electronic concentration camp in order to control the population? Is this part of the way in which modern governments will suppress "rogue nations" or their own populations? Is it possible that some other nation, corporation or individual might use these technologies to launch a "preemptive war" on the West?

158 "A Psy-Ops Bonanza On The Desert." ITV News Bureau, Ltd., 1991. http://www.mindspring.com/~silent/xx/daisy.htm . EPI568

159 ITV News Bureau, Ltd. "High Tech Psychological Warfare Arrives In The Middle East." 1991. http://www.mindspring.com/~silent/xx/news.htm EPI567

160. http://www.fas.org/man/dod-101/sys/ac/ec-130e.htm, Federation of American Scientists, Military Analysis Network. EPI6030

Back to the History

The development of the technology followed a very traceable history which began in the early 1960's at the height of the cold war. In 1961, Dr. Allen Frey wrote, "Our data to date indicates that the human auditory system can respond to electromagnetic energy in at least a portion of the radio frequency (RF) spectrum. Further, this response is instantaneous and occurs at low power densities, densities which are well below that necessary for biological damage. For example, the effect has been induced with power densities 1/60 of the standard maximum safe level for continuous exposure."[161] This observation had incredible ramifications because it meant that within certain ranges of RF people could create a sound perceived in the brain of a person at energy concentration levels considered too small to be significant.

Later that year a patent was issued to Dr. Puharich which stated in part, "The present invention is directed to a means for auxiliary hearing communication, useful for improving hearing, for example, and relates more specifically to novel and improved arrangements for auxiliary hearing communications by effecting the transmission of sound signals through the dental structure and facial nervous system of the user."[162] This crude device produced a signal which could be heard in the brain by inducing a vibration which was transferred through the bone into the inner ear where it was then carried to the brain via the nervous system. Puharich continued researching along this line, gaining an additional patent in 1965.[163] Both of these inventions required physical contact with the head of the subject.

161. Frey, Allan H. "Human Auditory System Response To Modulated Electromagnetic Energy." *Journal of Applied Physiology,* 17(4): 689-692. 1962. EPI544

162 US Patent #2,995,633, Aug. 8, 1961. Means for Aiding Hearing. Inventors: Puharich et al. EPI256

163. US Patent #3,170,993, Feb. 23, 1965. Means For Aiding Hearing By Electrical Stimulation Of The Facial Nerve System. Inventors: Puharich et al. EPI1119

By 1962, Dr. Allan Frey had advanced his work and was able to create sound at a distance from the subject using a pulsed (modulated) radio transmitter. "Using extremely low average power densities of electromagnetic energy, the perception of sounds was induced in normal and deaf humans. The effect was induced several hundred feet from the antenna the instant the transmitter was turned on, and is a function of carrier frequency and modulation."[164] What was occurring in this research were the first attempts to "tune" into the brain of a human in the same manner as "tuning" into a radio station. The same energy was being used, it was just at a different frequency with a slight vibration (pulse-modulation) on the carrier wave which delivered the signal. This system is wireless.

In 1968, a patent was issued for a device which also required physical contact with the skin of the test subject. "This invention relates to electromagnetic excitation of the nervous system of a mammal and pertains more particularly to a method and apparatus for exciting the nervous system of a person with electromagnetic waves that are capable of causing that person to become conscious of information conveyed by the electromagnetic waves."[165] This invention was much different than what others had created by that time because this device actually sent a clear audible signal through the nervous system to the brain. The device could be placed anywhere on the body and a clear voice or music would appear in the head of the subject. This was a most unbelievable device which had actually been invented in the late 1950's. It had taken years to convince patent examiners that it worked. The initial patent was only granted after the dramatic demonstration of the device on a deaf employee of

164 "Human Auditory System Response To Modulated Electromagnetic Energy", Allan H. Frey, *Journal of Applied Physiology*, 17(4): 689-692. 1962. EPI544

165. "Nervous System Excitation Device." US Patent No. 3,393,279, July 16, 1968, Assignee: Listening Incorporated. EPI261

the United States Patent Office. In 1972, a second patent was issued to the same researcher after being suppressed by the military for four years. This device was much more efficient in that it converted a speech waveform into "a constant amplitude square wave in which the transitions between the amplitude extremes are spaced so as to carry the speech information."[166] What this did is utilize the frequency code or timing sequences necessary for efficient transfers into the nervous system where the signals could be sent to the brain and decoded as sound in the same way that normal sound is decoded. The result was a clear and understandable sound.

In 2003, Robert Thiedemann of Munich, Germany, advanced the technology even further with the creation of the *Holophon®* in Germany and the *Earthpulse Soundwave*™ in the United States. This technology advanced the old design with higher fidelity and clarity in the sound transfer with a number of interesting effects being reported. The device was released in Europe and is scheduled for United States production beginning in early fall 2006. This device is used for positive applications by individuals and fully described in Part II of this book.

In 1971, a system was designed which would allow troops to communicate through a radio transmitter that rendered the enemy deaf and disoriented while allowing "friendly" combatants to communicate at the same time. The device was described as follows: "Broadly, this disclosure is directed to a system for producing aural and psychological disturbances and partial deafness of the enemy during combat situations. Essentially, a highly directional beam is radiated from a plurality of distinct transducers and is modulated by a noise, code, or speech beat signal. The invention may utilize various forms and may include movable radiators mounted on

166 "Method and System of Simplifying Speech Waveforms", U.S. Patent No. 3,647,970, March 7, 1972. EPI259

a vehicle and oriented to converge at a desired point, independently positioned vehicles with a common frequency modulator, or means employed to modulate the acoustical beam with respect to a fixed frequency. During combat, friendly forces would be equipped with a reference generator to provide aural demodulation of the projected signal, thereby yielding an intelligible beat signal while enemy personnel would be rendered partially deaf by the projected signal as well as being unable to perceive any intelligence transmitted in the form of a modulated beat signal."[167]What this says simply is that at-a-distance personal communication could be achieved by one's own forces while simultaneously denying it to adversaries using a wireless pulse-modulated signal.

In 1974, using a microwave, it was noted that the signal was changed (transduced) by the receiver into an acoustic signal. This was the signal that was "heard" inside or just behind the head. The report stated: "...it was noticed that the apparent locus of the 'sound' moved from the observer's head to the absorber. That is, the absorber acted as a transducer from microwave energy to an acoustic signal. This observation, to the best of our knowledge, has not been described in the literature and may serve as a mechanism mediating the 'hearing' of pulsed microwave signals."[168]

By 1989, the science took another leap forward with the combination of the modulated signal on a microwave carrier. This provided a much more efficient delivery of the sound. It was reported that, "Sound is induced in the head of a person by radiating the head with microwaves in the range of 100 megahertz to 10,000 megahertz that are modulated with a particular waveform. The waveform consists of frequency modulated bursts. Each burst is made up of ten to twenty

167 US Patent No.3,566,347, Feb. 23, 1971. "Psycho-Acoustic Projector". Inventor: Flanders, Andrew E. Assignee: General Dynamics Corporation. EPI260

168 "Generation of Acoustic Signals by Pulsed Microwave Energy."Sharp et al, *IEEE Transactions On Microwave Theory and Techniques*, May, 1974. EPI817

uniformly spaced pulses grouped tightly together. The burst width is between 500 nanoseconds and 100 microseconds. The pulse width is in the range of 10 nanoseconds to 1 microsecond. The bursts are frequency modulated by the audio input to create the sensation of hearing in the person whose head is irradiated."[169] This is the concept of the **frequency code** described in Chapter One of this book. Two patents were filed that year which addressed this breakthrough. The first "invention relates to devices for aiding of hearing in mammals. The invention is based upon perception of sounds which is experienced in the brain when the brain is subjected to certain microwave radiation signals."[170] And the second confirmed the earlier observations by stating that "Sound is induced in the head of a person by radiating the head with microwaves in the range of 100 megahertz to 10,000 megahertz that are modulated with a particular waveform. The waveform consists of frequency modulated bursts. Each burst is made up of ten to twenty uniformly spaced pulses grouped tightly together."[171]

In 1992, another patent described: "A silent communications system in which nonaural carriers, in the very low or very high audio frequency range or in the adjacent ultrasonic frequency spectrum, are amplitude or frequency modulated with the desired intelligence and propagated acoustically or vibrationally, for inducement into the brain, typically through the use of loudspeakers, earphones or piezoelectric transducers."[172] This device had limited practicality in that it required that the person be in contact or close proximity to the sending device. When examined together, each of these patents are seen to be discrete steps toward a new weapons system.

169 US Patent No.4,877,027, Oct. 31, 1989. Hearing System. Inventor: Wayne Brunkan. EPI1124

170 US Patent No.4,858,612, Aug. 22, 1989. Hearing Device. Inventor: William L.Stocklin. EPI270

171 US Patent No.4,877,027, Oct. 31, 1989. Hearing System. Inventor: Wayne B.Brunkan. EPI262

172 US Patent No. 5,159,703, Oct. 27, 1992. Silent Subliminal Presentation System. Inventor: Lowry, Oliver M. EPI285

These are the ways that the frequency codes of the human body are introduced in order to achieve the desired "controlled effect."

In 1995, it was reported that in the early research, clear sound signals had been sent and received. It is difficult now to determine what level of military or other research was being advanced in these areas. History was clear from Congressional Reports that this entire area was of great interest to the intelligence communities. "Drs. Allan Frey and Joseph Sharp conducted related research. Sharp himself took part in these experiments and reported that he heard and understood words transmitted in pulse-microwave analogs of the speakers sound vibrations. Commenting on these studies, Dr. Robert Becker, twice nominated for the Nobel Peace Prize, observed that such a device has obvious applications in covert operations designed to drive a target crazy with voices, or deliver undetectable instructions to a potential assassin."[173]

Later, in 1996 came the development of "A wireless communication system undetectable by radio frequency methods for converting audio signals, including human voice, to electronic signals in the ultrasonic frequency range, transmitting the ultrasonic signal by means of acoustic pressure waves across a carrier medium, including gases, liquids, or solids, and reconverting the ultrasonic acoustical pressure waves back to the original audio signal."[174] Although this was meant to be used with both receiving and sending hardware, what was determined were the modulation methods for transferring the signal. Wireless electronic telepathy had been invented.

173. "Nonlethal Defence: The New Age Mental War Zone." Issue 10, Scientists for Global Responsibility, 1995. EPI810
174. US Patent No. 5,539,705, July 23, 1996. Ultrasonic Speech Translator and Communication System. Inventor: Akerman, M. Alfred et al. Assignee: Martin Marietta Energy Systems. EPI293

The real work was yet to be made public in the form of patents. However, the military claims in the arena were starting to surface. What was known from experience was that patents were being held back by the government and confiscated by the military. When this intellectual property is seized the inventors are given a choice – work for the government or you cannot continue your research on or even talk about the invention under a national security order. Those who do not cooperate have their work effectively shut down.

Chapter Six

Other Patents
The Ethics of It All

Subliminal Messages and Commercial Uses

One of the areas where this new technology is being used is in systems to dissuade shoplifters, using sound below the range of hearing. "Japanese shopkeepers are playing CDs with subliminal messages to curb the impulses of the growing band of shoplifters. The mind control CDs have soundtracks of popular music or ocean waves, with encoded voices in seven languages, warning that anyone caught stealing will be reported to the police."[175] A number of patents have been developed to influence behavior in this way. The following summations are taken from some of these patents dealing with both audio and video programing, only this time, – *we are the program*:

> "An auditory subliminal programming system includes a subliminal message encoder that generates fixed frequency security tones and combines them with a subliminal message signal to produce an encoded subliminal message signal which is recorded on audio tape or the

175 "'Mind Control Music' Stops Shoplifters", Peter McGill, *The Sydney Morning Herald*, Feb. 4, 1995. EPI95

like. A corresponding subliminal decoder/mixer is connected as part of a user's conventional stereo system and receives as inputs an audio program selected by the user and the encoded subliminal message."[176]

"Ambient audio signals from the customer shopping area within a store are sensed and fed to a signal processing circuit that produces a control signal which varies with variations in the amplitude of the sensed audio signals. A control circuit adjusts the amplitude of an auditory subliminal anti-shoplifting message to increase with increasing amplitudes of sensed audio signals and decrease with decreasing amplitudes of sensed audio signals. This amplitude controlled subliminal message may be mixed with background music and transmitted to the shopping area."[177]

The idea of influencing the audio input to the brain in this way can also be manipulated using a visual stimulation. The next patents describe some of the concepts for accomplishing this kind of effect.

"Data to be displayed is combined with a composite video signal. The data is stored in memory in digital form. Each byte of data is read out in sequential fashion to determine: the recurrence display rate of the data according to the frame sync pulses of the video signal; the location of the data within the video image according to the line sync pulses of the video signal; and the location of the data display within the video image according to the position information."[178]

176 US Patent No.4,777,529, Oct. 11, 1988. Auditory Subliminal Programming System.Inventors: Schultz et al. Assignee: Richard M. Schultz and Associates, Inc. EPI265

177 US Patent No.4,395,600, July 26, 1983. Auditory Subliminal Message System and Method. Inventors: Lundy et al. EPI264

178 US Patent No.5,134,484, July 28, 1992. Superimposing Method and Apparatus Useful for Subliminal Messages. Inventor: Willson, Joseph. Assignee: MindsEye Educational Systems Inc. EPI290.

> "This invention is a combination of a subliminal message generator that is 100% user programmable for use with a television receiver. The subliminal message generator periodically displays user specified messages for the normal television signal for specific period of time. This permits an individual to employ a combination of subliminal and supraliminal therapy while watching television."[179]

It is too bad patent lawyers and inventors don't express themselves more clearly. Part of the "game" of patent filings is to give enough information to protect technology secrets while, at the same time, disclosing enough data to receive the patent based on the claims of the inventor. The patents may seem a bit complicated, however, they can be summarized. These patents are designed to provide a way to hide messages in video or audio formats masking any suggestions that the programmer wishes to convey. These kinds of messages bypass the conscious mind and are acted upon by the person hearing or seeing them – they are not sorted out by the active mind. Although these technologies are being developed for personal use and as security measures, consider the possibilities for abuse by commercial interests where the messages might be "buy, buy, buy," "drink more, don't worry, people who are promoted work 12 hours a day for free" or some other self-serving script. Should these systems be regulated? By whom and under what conditions?

New Standards for What is a Memory

"Nevada is currently the only state to allow witness testimony of a person who has undergone hypnosis. As of October 1, 1997, courts hearing both civil and criminal cases can take a hypnotically refreshed testimony, as long as the

179. US Patent No. 5,270,800, Dec. 14, 1993. Subliminal Message Generator. Inventor: Robert L. Sweet. EPI288

witness, if a minor, has had the informed consent of parent or guardian, and the person performing the hypnosis is any of the following: a health care provider, a clinical social worker licensed in accordance with 641B of the Nevada Revised statute, or a disinterested investigator."[180]

This issue will surely become more complex as technology advances in terms of evidence. If the day arrives when it is possible to completely change or alter memory as was suggested elsewhere by military officers, what then? How will we separate the real from the unreal? What will be the impact on the burden of proof in courts as it relates to "reasonable doubt"? Again the emergence of the technology has to first be recognized as real before laws can be constructed and systems established for controlling misuse. Consider how long it has taken the courts to even recognize hypnotherapy as valid science. We are hopeful that we will not have to wait so long for legislative bodies to take the initiative to address these issues.

Brain to Computer Connections

Major steps are being made to connect biology to information technology. The idea is centered on the notion that we can create any brain circuit and connect it to natural or grown biological materials. We can create brain cells, and the neurons that connect them, and then link these to electronic circuits.

As biological science, information technologies, communications systems and miniaturization of electronic designs (down to a one-billionth scale in size) advance soon mankind will be able to equip people with technologies that

180"Watch Carefully Now: Solving Crime in the 21st Century." E. Gene Hall, *Police*, June 1999. Vol. 23, No. 6. Source: NLECTC *Law Enforcement & Technology News Summary*, June 17, 1999. EPI944

replace, and may even significantly increase, our natural capacities. "Researchers said they took a key first step toward creating electronic microchips that use living brain cells. The researchers said they had learned how to place embryonic brain cells in desired spots on silicon or glass chips and then induce the brain cells to grow along desired paths."[181] In addition, "Scientists have succeeded for the first time in establishing a colony of human brain cells that divide and grow in laboratory dishes, an achievement with profound implications for understanding and treating a wide range of neurological disorders from epilepsy to Alzheimer's disease."[182] The other possibility is that both brain cells and computer hardware could be built in the laboratories creating, perhaps, the first biologically augmented computers.

How Powerful is the Brain?

In a recent report[183] Japanese automaker Honda, in cooperation with ATR Computational Neuroscience Laboratories, announced the development of a system that reads brain patterns but without any implants or surgery. Brain activity is scanned and its codes changed into signals that drive digital devices and robotic arms. This is a way that the combined technologies will be used to map and control functions. As a device or technology for replacing cell phones, keyboards and damaged portions of the brain a great deal can, and will, come from this type of research.

The idea of "blueprinting" the human brain is gaining strength and possibilities according to one story reported on CNet News. Using an IBM Blue Gene/L Supercomputer the

181 "Nervy Scientists Move Toward Union of Living Brain Cells with Microchips." Jerry Bishop, *Wall Street Journal*, Feb. 1, 1994, pp B3. EPI49

182 "Scientists Make Brain Cells Grow", Michael Specter, *Anchorage Daily News*, May 4, 1990. EPI527

183. "New Technology Uses Brain Signals to Control Robot", *Associated Press* , May 25, 2006, Tokyo, EPI6066

Ecole Polytechnique Federale de Lausanne (EPFL) in Switzerland was able to conduct a three dimensional simulation of 10,000 neurons firing in the brain. This required a terabyte of data which is a fraction of what would be needed to map the brain's billions of neurons and reduce the information to known algorithms.[184] The Swiss are also trying to network three of these computers together in the hope of increasing the ability to record and further analyze more complex brain activities. Those billions of neurons can be reduced to millions with respect to very specific areas of the brain responsible for emotions or other specific brain functions. Using the full capacity of the computers, networked together, might increase this to simulations of 5,000,000 neurons at optimum speeds of 280.6 terraflops per machine.

To understand and visualize this amount of computing power we have to consider what an IBM Blue Gene/L Supercomputer is really. The IBM Blue Gene/L Supercomputer at the Lawrence Livermore National Laboratory in California, as of June 30th, 2006 was the fastest computer in the world able to process at speeds of 280.6 terraflops per second (It would take sixty hours for the 6.6 billion people on the planet, using hand calculators, performing a calculation every five seconds, to accomplish what this supercomputer can do in 1 second).[185] To put this in the proper perspective we have to keep in mind how fast computers are increasing in power. According to Moore's Law, computer chip capacity is doubling every eighteen months, while supercomputing power is doubling from the previous year every eleven to twelve months.[186] We are only a few years away from full brain mapping with existing technology evolution speeds.

184. "Blueprinting the Human Brain", by Stefanie Olsen, Staff Writer, *CNet News.com,* May 10, 2006, EPI6076
185. "Top Computer Hangs on to Its Title", By Ned Stafford, *news@nature.com The Best in Science Journalism,* June 30, 2006 EPI6077
186. Ibid.

On the other side of the issue is a better understanding of exactly how powerful the human brain truly is and what it is capable of doing. The idea of enhancing human performance, based on what we know, could lead to individuals able to consciously do with their brains, in one second, what 6.6 billion people cannot accomplish if they all had hand calculators running for decades.

What's on Your Mind?

There are a number of uses being considered for technologies that have an impact on mental processes. There is an increasing call for research in areas that will permit brain mapping and then the generation of information that can be understood by computers so that a person's thoughts can be recorded, understood and reprogrammed at various levels. The concept of influencing people through electronic telepathy and mind reading is possible and becoming even more refined as these technologies advance. Although the idea of telepathy may become understood through the development of these technologies it is already perhaps an anomalous capacity all people have if we develop it. The idea of changing brain activity, even our own thoughts, is the stuff of modern science and no longer the imaginings of science fiction writers.

A significant initiative was started for use in creating counter drug measures, the "Brain Imaging Technology Initiative. This initiative establishes NIDA regional neuro-imaging centers and represents an interagency cooperative endeavor funded by CTAC, Department of Energy (DOE), and NIDA to develop new scientific tools (new radiotracers and technologies) for understanding the mechanisms of addiction and for the evaluation of new pharmacological treatments."[187] Through neuroimaging, not only could the stated objective be

187 ONDCP,CTAC. Counterdrug Research and Development Blueprint Update: http://www.whitehousedrugpolicy.gov/scimed/ blueprint99/execsumm.htm . EPI305

achieved, but through imaging, a person's emotional states could be mapped, chemical influences determined and perhaps even specific thoughts could be read.

In 1975: "Developments in ways to measure the extremely weak magnetic fields emanating from organs such as the heart, brain and lungs are leading to important new methods for diagnosing abnormal conditions."[188] This knowledge is used to interpret and apply the frequency codes, using magnetic fields, as a way to facilitate induction, they can affect specific targeted body organs.

In 1995 a system for capturing and decoding brain signals was developed which includes a transducer for stimulating a person's brain; EEG transducers for recording brain wave signals from the person; a computer for controlling and synchronizing stimuli presented to the person. At the same time the system could record brain wave signals, and either interpret signals using a model for conceptual, perceptual and emotional thought to correspond to the EEG signals of thought of the person, or compare the signals to normal brain signals from a normal population to diagnose and locate the origin of brain dysfunction underlying perception, conception and emotion.[189] In other words, the device reads one's mind by comparing your brain activity to other people's and then by using this information they can create new signals and send them back to change the perception or emotion.

In 1996, came this Orwellian development, "a method for remotely determining information relating to a person's emotional state, as waveform energy having a predetermined frequency and a predetermined intensity is generated and

188."Magnetic Fields of the Human Body", David Cohen, *Physics Today*, August, 1975. EPI1179

189 US Patent No. 5,392,788, Feb. 28, 1995. Method and Device for Interpreting Concepts and Conceptual Thought from Brainwave Data and for Assisting for Diagnosis of Brainwave Dysfunction. Inventor: William J.Hudspeth. EPI1129

wirelessly transmitted towards a remotely located subject. Waveform energy emitted from the subject is detected and automatically analyzed to derive information relating to the individual's emotional state. Physiological or physical parameters of blood pressure, pulse rate, pupil size, respiration rate and perspiration level are measured and compared with reference values to provide information utilizable in evaluating interviewee's responses or possibly criminal intent in security sensitive areas."[190] This technology could be used for determining what a person might do, given his totally discernible interior emotions. This technology walks through any behavior wall a person can erect and goes straight to the brain to see what might be on a person's mind.

In 1991, a method for changing brain waves to a desired frequency was patented.[191] A 1975 patent discussed a similar technology, a device and method for "sensing brain waves at a position remote from a subject whereby electromagnetic signals of different frequencies are simultaneously transmitted to the brain of the subject in which the signals interfere with one another to yield a waveform which is modulated by the subject's brain waves. The interference waveform which is representative of the brain wave activity is retransmitted by the brain to a receiver where it is demodulated and amplified. The demodulated waveform is then displayed for visual viewing and then routed to a computer for further processing and analysis. The demodulated waveform also can be used to produce a compensating signal which is transmitted back to the brain to effect a desired change in electrical activity therein."[192] In simple terms, the brain's activity is mapped in

190 US Patent No. 5,507,291, April 16, 1996. Method and an Associated Apparatus for Remotely Determining Information as to Person's Emotional State, Inventors: Stirbl et al. EPI1130

191 US Patent No. 5,036,858, Aug. 6, 1991. Method And Apparatus For Changing Brain Wave Frequency. Inventors: Carter et al. EPI1127

192 US Patent No. 3,951,134, April 20, 1976. Apparatus And Method For Remotely Monitoring And Altering Brainwaves. Inventor: Robert G. Malech, Assignee: Dorne & Margolin, Inc. EPI1122

order to read a person's emotional state, conceptual abilities or intellectual patterns. A second signal can be generated and sent back into the brain which overrides the natural signal, causing the brain's energy patterns to shift. This is called "brain entrainment" which causes the shift in consciousness. This can also develop into a direct memory transfer technology. There are many uses of a positive nature for this kind of technology; as is mentioned elsewhere, the important factor being who controls the technology and for what purposes. In the leading scientific journal, *Nature,* the following encapsulating statement appeared:

> "'But neuroscience also poses potential risks', he said, arguing that advances in cerebral imaging make the scope for invasion of privacy immense...it will become commonplace and capable of being used at a distance, he predicted. That will open the way for abuses such as invasion of personal liberty, control of behavior and brainwashing."[193]

Still Wondering What's on Your Mind

In "a dramatic demonstration of mind reading, neuroscientists have created videos of what a cat sees by using electrodes implanted in the animal's brain. 'Trying to understand how the brain codes information leads to the possibility of replacing parts of the nervous system with an artificial device,' he said."[194] The scientist commenting on this technology saw the future possibility of brain activity mapping being used in creating electronic components to replace damaged parts of the brain. The use of mind mapping has other possibilities as well.

193 *Nature.* "Advances in Neuroscience 'May Threaten Human Rights.'" Vol. 391, Jan. 22, 1998. EPI116

194. "A Cat's Eye Marvel", Kahney, Leander, *Wired News,* Oct. 7,1999. http://www.wired.com/news/technology/story/22116.html EPI832

By 1998, publicly released information was being discussed as a result of information openly flowing out of Russia. A meeting was held to assess the threat. The "main purpose of the March meetings was described in the Psychotechnologies memo to 'determine whether psycho-correction technologies represent a present or future threat to U. S. national security in situations where inaudible commands might be used to alter behavior.'"[195] The threat assessment was likely to begin to condition Americans for the public acknowledgment of one of the government's long held secrets – the human mind and body could be controlled remotely without a trace of evidence being left behind. In another quote one of the leading researchers in this area began to announce his findings: "But psychological-warfare experts on all sides still dream that they will one day control the enemy's mind. And in a tiny, dungeon like lab in the basement of Moscow's ominously named Institute of Psycho-Correction, Smirnov and other Russian psychiatrists are already working on schizophrenics, drug addicts and cancer patients."[196] The results of this research have been investigated and demonstrated to members of the intelligence community in the United States and have even been demonstrated on a Canadian television documentary by Dr. Smirnov. In the same TV segment where I appeared and demonstrated an infrasound device, Dr. Smirnov demonstrated his technology.[197]

In a more recent article in *Pravda,*[198] Dr. Smirnov is referred to as the father of psychotropic weapons in Russia. He was the first to experiment with psycho-reconnaissance in 1984, and was involved throughout his career in developing mind control technologies. Smirnov discusses in the context of

195. "DoD, Intel Agencies Look at Russian Mind Control Technology, Claims." *Defense Electronics,* July, 1993. EPI538

196 "A Subliminal Dr. Strangelove ", by Dorinda Elliot and John Barry, *Newsweek,* Aug. 21, 1994. EPI542

197. *Undercurrents,* CBC-TV, Canada, February, 1999.

198. Mind Control: The Zombie Effect, *Pravda,* November 10, 2004. EPI6039

the article how they too have been able to create a "Manchurian Candidate" by splitting a personality. He said, "As a result of such 'outside influences', a person's 'self' gets totally blocked. Instead, another 'self' is being created. That second identity in turn can have a number of various programmed urges, such as killing oneself." He discusses the use of masked subliminals and other means for by-passing the conscious mind and directly influencing the subconscious.

"Fantasies are thought processes involving internal monologues and imaginative sequences which can motivate healthy people to constructive behavior; likewise, they can inspire unbalanced individuals to destructive or dangerous behavior. One conclusion from that research was that fantasy played a major role among violent criminals. Researchers learned that criminals often daydreamed their fantasies, and then practiced elements of those fantasies before committing their crime. FBI agents determined that violent criminals often exhibit telltale signs as children and as adults. Hence, disturbed employees or students may demonstrate signs of violent fantasies to close observers. Troubled individuals may be obsessively interested in music with violent lyrics, or may have a drug or alcohol problem. When these signs reveal themselves, they should be reported to a threat management team, which can then neutralize the threat, either by therapy, if rehabilitation is possible, or by firing the employee. Workplace and school violence is usually preceded by warning signs."[199]

The ability to determine a "predisposition" for a behavior does not mean that a person will make the "choice" to act on the feelings and internal thoughts. Every person on the planet can remember times when his thoughts were dangerous, immoral or otherwise unacceptable, falling below

---

199 "To Dream, Perchance to Kill." Roger L. Depue and Joanne M. Depue, *Security Management*, July,1999. Vol. 43, No. 6. Source: NLECTC *Law Enforcement & Technology News Summary*, July 8, 1999. EPI932

the standards set by societal and cultural "norms." Yet, we could have these thoughts in the privacy of our own mind. The trend in the uses of mind control technology now would make our most private internal thoughts, as we wrestle with the temptations and choices of everyday life, subject to scrutiny by government and employers. Who will define the rules for psychocorrection? Who will decide what is ethical and right in this area as it develops over the next decade?

The American Civil Liberties Union (ACLU) on June 26, 2006, filed a Freedom of Information Request[200] with the Department of Defense (DoD), Defense Intelligence Agency (DIA), National Security Agency (NSA), Central Intelligence Agency (CIA) and the Federal Bureau of Investigation (FBI) seeking information on "the use of functional magnetic resonance imaging (fMIR), electroencephalography (EEG), infrared spectroscopy, or other scanning and measurement technologies of the brain that seek to detect the truth, deception, guilty knowledge, accurate recollections or recognition, or to otherwise assist in or support interrogations or to identify individuals for follow-up questioning." Specifically, they sought "all records including but not limited to the study, development or use of such technologies for foreign or domestic use and all records of the agency's efforts to contract with any other entity to study, develop or use such technologies for foreign or domestic use." The main argument of the ACLU was about the "effectiveness" of the technologies[201] being used. I am sure, if the ACLU ever receives a complete answer to their request, through this administration's fog of misinformation, it will be an eye-opener of what is really being done in the development of these technologies.

200. Freedom of Information Request, filed by the ACLU Technology and Liberty Program, June 26,2006. EPI 6078
201. "ACLU Seeks Information about Government Use of Brain Scanners in Interrogations, June 28, 2006. EPI6079

Chapter Seven

Control of the Mind and Body

The predominant brain wave frequencies indicate the kind of activity taking place in the brain. There are four basic groups of brain wave frequencies which are associated with most mental activity. The first, beta waves, (13-35 Hertz or pulses per second) are associated with normal activity. The high end of this range is associated with stress or agitated states which can impair thinking and reasoning skills. The second group, alpha waves (8-12 Hertz), can indicate relaxation. Alpha frequencies are ideal for learning and focused mental functioning. The third, theta waves (4-7 Hertz), indicate mental imagery, access to memories and internal mental focus. This state is often associated with young children, behavioral modification and sleep/dream states. The last, ultra slow, delta waves (.5-3 Hertz), are found when a person is in deep sleep. The general rule is that the brain's predominant wave frequency will be lowest, in terms of pulses per second, when relaxed, and highest when people are most alert or agitated.[202]

External stimulation of the brain by electromagnetic means can cause the brain to be entrained or locked into phase with an external signal generator.[203] Predominant brain waves

202. *Mega Brain, New Tools and Techniques for Brain Growth and Mind Expansion*, by Michael Hutchison, 1986. EPI1235

203 US Patent No. 5,356,368, Oct. 18, 1994. Method and Apparatus for Inducing Desired States of Consciousness. Inventor: Monroe, Robert. Assignee: Interstate Industries, Inc. EPI286

can be driven or pushed into new frequency patterns by external stimulation. In other words, the external signal driver or impulse generator entrains the brain, overriding the normal frequency codes, and causing changes in the brain waves; which then cause changes in brain chemistry; which then cause changes in brain outputs in the form of thoughts, emotions or physical conditions. As you are driven, so you arrive – brain manipulation can be either beneficial or detrimental to the individual being impacted, depending on the level of knowledge or the intentions of the person controlling the technology.

In combination with specific wave forms, the various frequencies trigger precise chemical responses in the brain. The release of these neurochemicals causes specific reactions in the brain which result in feelings of fear, lust, depression, love, etc. All of these, and the full range of emotional/intellectual responses, are caused by very specific combinations of these brain chemicals which are released by frequency-specific electrical impulses. "Precise mixtures of these brain juices can produce extraordinarily specific mental states, such as fear of the dark, or intense concentration."[204] The work in this area is advancing at a very rapid rate with new discoveries being made regularly. Unlocking the knowledge of these specific frequency code is yielding significant breakthroughs in understanding human health.

The control of mind and body using various forms of electromagnetic energy, including radio signals, light pulsations, sound and other methods has resulted in several inventions and innovations. The positive health effects and uses have been pursued by private researchers around the world. In 1973, an "apparatus for the treatment of neuropsychic and somatic disorders wherein light, sound, VHF electromagnetic field and heat sources, respectively, are simultaneously applied

204. *Mega Brain, New Tools and Techniques for Brain Growth and Mind Expansion*, by Michael Hutchison, 1986, pp 114. EPI1235

by means of a control unit to the patient's central nervous system with a predetermined repetition rate was invented. The light radiation and sound radiation sources are made so as to exert an adequate and monotonous influence of the light-and-sound-radiation on the patient's visual analyzers and auditory analyzers, respectively."[205] This results in the brain following the external simulating source in triggering brain pattern changes which effect the brain immediately and directly. This is again a coherent signal that man can create and the human brain can understand and respond to in a manner that cannot be resisted. It is the frequency following response (FFR) that results.

In 1980, another patent was issued which disclosed "a method and apparatus for producing a noise-like signal for inducing a hypnotic or anesthetic effect in a human being. The invention also has uses in crowd control and consciousness level training (biofeedback). The invention may also be used in creating special musical effects."[206] This device would have a profound effect in controlling individuals to a point otherwise only achievable through the application of hypnotherapy or drugs. This type of controlled effect can be induced from a distance, outside of the human body, without implants or physical contact of any kind.

A couple of years later another device was engineered to create these types of effects, again using very subtle energy resulting in: "Brain wave patterns associated with relaxed and meditative states in a subject are gradually induced without deleterious chemical or neurological side effects."[207] Again

---

205 US Patent No. 3,773,049, Nov. 20, 1973. Apparatus for the Treatment of Neuropsychic and Somatic Diseases with Heat, Light, Sound and VHF Electromagnetic Radiation. Inventors: Rabichev, et al. EPI257

206 US Patent No. 4,191,175, March 4, 1980. Method and Apparatus for Repetitively Producing a Noise-Like Audible Signal. Inventor: William L. Nagle. EPI269

207 US Patent No. 4,335,710, June 22, 1982. Device For the Induction of Specific Brain Wave Patterns. Inventor: John D. Williamson, Assignee: Omnitronics Research Corporation. EPI292

changing our chemistry by the use of an external energy source.

Various systems were perfected and patents were eventually issued for a number of innovative technologies for controlling brain activity in people.[208,209,210,211,212,213,214,215] These inventions generated a whole array of breakthroughs for controlling a person's emotional state, concentration, and pain levels, and creating other effects as well. In 1990, the results of a study strongly indicated "that specific types of subjective experiences can be enhanced when extremely low-frequency magnetic fields of less than 1 milligauss are generated through the brain at the level of the temporal lobes. Vestibular feelings (vibrations, floating), depersonalization (feeling detached, sense of a presence) and imaginings (vivid images from childhood) were more frequent within the field exposed groups than the sham-field exposed group."[216]The results were that a complete array of research projects over decades were beginning to yield incredible combinations and results.

208 US Patent No. 4,717,343, Jan. 5, 1988. Method of Changing a Person's Behavior. Inventor: Alan Densky. EPI284

209 US Patent No. 4,834,701, May 30, 1989. Apparatus for Inducing Frequency Reduction in Brainwaves, Inventor: Masaki, Kazumi. Assignee: Ken Hayashibara. EPI266

210 US Patent No. 4,889,526, Dec 26, 1989. Non-Invasive Method and Apparatus for Modulating Brain Signals Through an External Magnetic or Electric Field to Reduce Pain. Inventors, Elizabeth A. Rauscher and William L. Van Bise, Assignee: Megatech Laboratories, Inc. EPI268

211 US Patent No. 4,227,516, Oct. 14, 1990. Apparatus for Electrophysiological Stimulation. Inventors: Meland, et al. EPI283

212 US Patent No. 4,883,067, Nov. 28, 1989. Method and Apparatus for Translating the EEG into Music to Induce and Control Various Psychological and Physiological States and to Control a Musical Instrument. Inventors: Knispel, et al. Assignee: Neurosonics, Inc. EPI282

213 US Patent No. 5,123,899, June 23, 1992. Method and System for Altering Consciousness. Inventor: Gall, James. EPI289.

214 Patent No. 5,352,181, Oct. 4, 1994. Method and Recording for Producing Sounds and Messages to Achieve Alpha and Theta Brainwave States and Positive Emotional States in Humans. Inventor: Mark E. Davis. EPI291

215 US Patent No. 5,289,438, Feb. 22, 1994. Method and System for Altering Consciousness. James Gall. EPI333

216 "Enhancement of Temporal Lobe-Related Experiences During Brief Exposures to Milligauss Intensity Extremely Low Frequency Magnetic Fields.", Leslie A. Ruttan, et al.*Journal of Bioelectricity*, 9(1), 33-54 (1990). EPI311

In another 1996 new age sounding invention, quartz crystals are used to create stress relief by slowing brain activity. "Physiological stress in a human subject is treated by generating a weak electromagnetic field about a quartz crystal. The crystal is stimulated by applying electrical pulses of pulse widths between 0.1 and 50 microseconds each, at a pulse repetition rate of between 0.5k and 10k pulses per second, to a conductor positioned adjacent to the quartz crystal, thereby generating a weak magnetic field. A subject is positioned within the weak magnetic field for a period of time sufficient to reduce stress."[217] The crystal is piezoelectric and tunes the signal.

Consciousness training is also a big theme in cults, religious organizations and others pursuing the "new age". Science has now gained a greater understanding of how the mind and brain work, so that what used to take years, or even decades to achieve, can now be mastered in weeks, days or even minutes. For instance, in 1996, "a method and apparatus for use in achieving alpha and theta brain wave states and effecting positive emotional states in humans,"[218] was developed. Two years later another patent was issued which could create desired consciousness states: in the training of an individual to replicate such states of consciousness without further audio stimulation; and in the transferring of such states from one human being to another through the imposition of one person's EEG, superimposed on desired stereo signals, on another individual, by inducement of a binaural beat phenomenon."[219] Thought transference? This is interesting in that it speaks to the ideas alluded to earlier by the military in

217. US Patent No. 5,562,597, Oct. 8, 1996. Method and Apparatus for Reducing Physiological Stress. Inventor: Robert C. Van Dick. EPI294

218 US Patent No. 5,586,967, Dec. 24, 1996. Method and Recording or Producing Sounds and Messages To Achieve Alpha and Theta Brainwave States in Humans. Inventor: Mark E. Davis. EPI296

219 US Patent No. 5,213,562, May 25, 1993. Method of Inducing Mental, Emotional and Physical States of Consciousness, Including Specific Mental Activity in Human Beings. Inventor: Robert Monroe, Assignee: Interstate Industries, Inc. EPI287

changing the memory of a person by imposing computer manipulated signals which would integrate with the normal memory of a person. The possibilities of abuse are obvious and the opportunity for personal advancement is also great. Consider the possibility of gaining education by the transfer of data directly into the human brain by these new methods rather than the standard methods of learning. A serious consideration in developing these types of memory transfer systems will be the fact that they bypass normal intellectual filters – they are deposited into the brain as fact, without question or careful consideration. What happens when the new information conflicts with existing information? Would it be possible to include hidden information meant to unduly influence things like religious beliefs, politics or consumption of goods and services, etc.?

The possibilities are immense and the ethical and moral questions surrounding these matters are equally significant. We can no longer avoid the debate, in fact the debate is lagging far behind the scientific advancements. In the interim, there are some simple things we could all do to enhance our own, or our children's, learning capacity by applying simple and available knowledge. For instance: "researchers at the Center for the Neurobiology of Learning and Memory at the University of California, Irvine, have determined that 10 minutes of listening to a Mozart piano sonata raised the measurable IQ of college students by up to nine points."[220] This is a simple thing of great use to anyone seeking self improvement. There are *many* more uses of a positive nature that are discussed throughout this book.

220 "Listening to Mozart a Real - But Temporary - IQ Builder, Study Says.", by Robert Lee Hotz, *Anchorage Daily News*, Oct. 15, 1993. EPI529

Chapter Eight

Weaponization of the Mind

The idea of influencing and controlling the mind and body, by external means, has been of great interest to the military. The impact of the science is just now being recognized although referenced in the open literature. The below material was put together and used to press the research forward in terms of influencing policy within the defense department.

> "The results of many studies that have been published in the last few years indicate that specific biological effects can be achieved by controlling the various parameters of the electromagnetic (EM) field. A few of the more important EM factors that can be manipulated are frequency, wave shape, rate of pulse onset, pulse duration, pulse amplitude, repetition rate, secondary modulation, and symmetry and asymmetry of the pulse. Many of the clinical effects of electromagnetic radiation were first noticed using direct current applied directly to the skin. Later the same effects were obtained by applying external fields. Electromagnetic radiation has been reported in the literature to induce or enhance the following effects:

1. Stimulation of bone regeneration in fractures.
2. Healing of normal fractures.
3. Treatment of congenital pseudarthosis.
4. Healing of wounds.
5. Electroanesthesia.
6. Electroconvulsive therapy.
7. Behavior modification in animals.
8. Altered electroencephalograms in animals and humans.
9. Altered brain morphology in animals.
10. Effects of acupuncture.
11. Treatment of drug addiction.
12. Electrostimulation for relief of pain.

These are but a few of the many biological effects and uses that have been reported over the past decade. They are not exhaustive, and they do not include many of the effects reported in the Soviet and East European literature.

As with most human endeavors, these applications of electromagnetic radiation have the potential for being a double-edged sword. Electromagnetic radiation applications can produce significant benefits, yet at the same time can be exploited and used in a controlled manner for military or certain covert uses. This paper focuses on the potential uses of electromagnetic radiation in future low-intensity conflicts.

POTENTIAL MILITARY APPLICATIONS OF EMR

The exploitation of this technology for military uses is still in its infancy and only recently has been recognized by the United States as a feasible option. A 1982 Air Force review of biotechnology had this to say:

'Currently available data allow the projection that specially generated radio frequency radiation (RFR) fields may pose powerful and revolutionary antipersonnel military threats. Electroshock therapy indicates the ability of induced electric current to completely interrupt mental functioning for short periods of time, to obtain cognition for longer periods and to restructure emotional response over prolonged intervals.

Experience with electroshock therapy, RFR experiments and the increasing understanding of the brain as an electrically mediated organ suggested the serious probability that impressed electromagnetic fields can be disruptive to purposeful behavior and may be capable of directing and or interrogating such behavior. Further, the passage of approximately 100 milli-amperes through the myocardium can lead to cardiac standstill and death, again pointing to a speed-of-light weapons effect.

A rapidly scanning RFR system could provide an effective stun or kill capability over a large area. System effectiveness will be a function of wave form, field intensity, pulse widths, repetition frequency, and carrier frequency. The system can be developed using tissue and whole animal experimental studies, coupled with mechanisms and waveform effects research.

Using relatively low-level RFR, it may be possible to sensitize large military groups to extremely dispersed amounts of biological or chemical agents to which the unirradiated population would be immune.

The potential applications of artificial electromagnetic fields are wide ranging and can be used in many military or quasi-military operations.

> Some of these potential uses include dealing with terrorist groups, crowd control, controlling breeches of security at military installations, and antipersonnel techniques in tactical warfare...One last area where electromagnetic radiation may prove to be of some value is in enhancing abilities of individuals for anomalous phenomena."[221]

Quite the paper for 1984. Stimulating anomalous phenomena was another interesting point revealed in this paper. What could this mean? In one press report the interest of the CIA was disclosed when it was announced that for "20 years, the United States has secretly used psychics in attempts to help drug enforcement agencies track down Libyan leader Moammar Gadhafi and find plutonium in North Korea, the CIA and others confirm. The ESP spying operations – code named 'Stargate' – were unreliable, but three psychics continued to work out of Ft. Meade, at least into July, researchers who evaluated the program for the CIA said."[222] It is also worth pointing out that this report coincided with the public disclosure by military personnel of this project. The story was revealed in a book called *Psychic Warrior*. John Alexander, working out of Los Alamos, and a major proponent of this area of research, wrote that "there are weapons systems that operate on the power of the mind and whose lethal capacity has already been demonstrated...The psychotronic weapon would be silent, difficult to detect, and would require only a human operator as a power source."[223]

221. *Low-Intensity Conflict and Modern Technology,* Lt. Col. David Dean, Air University Press, June, 1986. EPI709
222 "ESP Spies, 'Stargate' are Psychic Reality", by Richard Cole, *Saint Paul Pioneer Press,* Nov. 30, 1995. EPI491
223 "The "Soft Kill" Fallacy", by Steven Aftergood, *Bulletin of the Atomic Scientists,* September/October , 1994. EPI281

Other Exotic Technologies

An "RF weapon currently under development is the high-powered, very low frequency (VLF) modulator. Working in the 20-35 KHz spectrum, the frequency emits from a 1-2 meter antenna dish to form in a type of acoustic bullet. The weapon is especially convenient because the power level is easily adjustable. At its low setting, the acoustic bullet causes physical discomfort – enough to deter most approaching threats. Incrementally increasing the power nets an effect of nausea, vomiting and abdominal pains. The highest settings can cause a person's bones to literally explode internally. Aimed at the head, the resonating skull bones have caused people to hear 'voices'. Researched by the Russian military more extensively than by the U.S., the Russians actually offered the use of such a weapon to the FBI in the Branch Davidian standoff to make them think that 'God' was talking to them. Concerned with the unpredictability of what the voices might actually say to the followers, the FBI declined the offer. Another RF weapon that was ready for use back in 1978 was developed under the guise of Operation PIQUE. Developed by the CIA, the plan was to bounce high powered radio signals off the ionosphere to affect the mental functions of people in selected areas, including Eastern European nuclear installations."[224]

Back to Dr. Jose Delgado

The effects that Delgado's teams in Spain were studying included behavior modification in animals, inhibiting bacterial growth, altering migratory movements in snails, and healing bone fractures in rats. Electromagnetic fields (EMF) have also produced some effects with significant negative ramifications, including genetic mutations in fruit flies and gross deformations

224. "Nonlethal Weapons for Military Operations other Than War.", Joseph M.Suhajda, http://www.usafa.af.mil/wing/34edg/airman/suhajd~1.htm EPI348

in chick embryos.[225, 226] What this implies is that these low level effects alter the building blocks of the cells (the genetic code), which then alter the body components, which then alter the body. This translates into big problems for people when applied to human physiology. Scientists have known for some time that high levels of this type of radiation (ionizing radiation) can heat up the body, but they had assumed that low levels were not a problem. However, the current research, including military research, indicates that this is not the case. "For better or worse, weak electromagnetic fields are emerging as a potent biological force – a discovery with staggering implications for future medicine, science, industry and perhaps even the military... There is growing concern among experts that EMFs have been – or will be – deployed as invisible weapons to disrupt brain functioning and health."[227]

The work of Delgado would not be particularly alarming if it were isolated. However, "animals exposed to radiation from radio waves show subtle changes in blood cell count, immunity, the nervous system, and behavior, according to a 1984 EPA document."[228] The majority of these observed effects were from exposures which were within the government approved safety thresholds.

The idea of military planners, using the technology to control rather than enhance human performance, is often the fundamental root of their research. There are major opportunities that are missed when the greater emphasis is placed on military rather than wider uses of emerging science. The advances in these areas parallel many others, but lag behind in the application of ethics in the context of their

225. "The Mind Fields", by Kathleen McAuliffe, *Omni Magazine* Feb1985
226. *Cross Currents, The Perils of Electropollution, The Promise of Electromedicine*, by Robert O. Becker, M.D., pg 210.
227. "The Mind Fields", by Kathleen McAuliffe, *Omni Magazine*, February 1985.
228. Ibid.

development. An integrated discussion of the applied technology, civil liberties, ethics and private uses must take place. Consideration needs to be given to what should migrate to the private sector for use in enabling enhancement of human potentials in ways only reserved in the past to mystics, religious figures and those who sought to change people. The understanding of these technologies can serve to unleash our "anomalous human potentials" that are only rarely demonstrated. There is a great level of responsibility for the outcomes when a technology is used, with or without a person's consent, to change their consciousness, physical health or mental capabilities.

Chapter Nine

For Whose Use? The New Tools of Control

The Nonlethal Conference

In 1986, the United States' Attorney General held a conference on the development of these new high-tech weapons. It was summarized the following year[229] in a report discussing problems and solutions for what was believed to be an area with significant future possibilities. Several areas were reported to offer new possibilities for the Department of Justice.

The Department of Justice's report disclosed that:

"Participants also discussed the use of various wave lengths and forms of administration of electromagnetic energy as a non-lethal weapon. A substantial amount of preliminary research has been conducted in this area... One conference participant noted that scientific knowledge of human physiology is progressing to the point where it may soon be possible to target specific physiologic systems with specific frequencies of electromagnetic radiation to produce much more subtle and

229. *Report on the Attorney General's Conference on Less Than Lethal Weapons*, by Sherri Sweetman, March,1987, U.S. Department of Justice, National Institute of Justice.

fine-tuned effects than those produced by photic driving."

Photic driving is the use of visible light pulsed at frequencies which have been shown to impact the brain. In some people they cause epileptic seizures. The report went on:

"There is some evidence (and a good deal of supposition) that sustained, extremely low frequency (ELF) radiation can produce nausea or disorientation. One researcher has subjected animals to ELF electromagnetic radiation through electrode implants, and feels that similar results could be produced from afar, without electrodes. One participant suggested that ideally, one might like to develop the ability to design these electromagnetic fields for specialized use, for instance to produce sleep or confusion. It is known that sleep can be induced by electrodes in the brain, and Russian scientists claim to be able to produce sleep from afar (electrosleep)."[230]

What these conference participants were referring to was the work conducted by Dr. Jose Delgado of Yale University and the Russian LIDA machine tested by Dr. Adey, both referenced earlier. Many of the previously cited patents explain how this has since been achieved.

Los Alamos and Friends

The report on the Department of Justice's 1986 conference was also noted, "The nature of weapons to be tested and the necessity for the tests must not be a secret of the kind whose 'leak' would result in an exaggerated expose and associated public outcry." The report went further to

230. *Report on the Attorney General's Conference on Less Than Lethal Weapons*, by Sherri Sweetman, March,1987, U.S. Department of Justice, National Institute of Justice.

recognize that the military had made a number of advances in these areas but that these advances were classified. It also recommended that the military should develop the technologies for less-than-lethal weapons, because they were technically sophisticated enough to do the job and the Department of Justice was not set up for this kind of research initiative.

The concern about open investigation must have been laid aside when the true extent of the military advancements were eventually realized. The idea of an open discussion of these emerging areas was abandoned to the shadowy world of secrecy and hidden agendas. The next major conference where the military unveiled its new technologies to the Justice Department was classified and not open to the press or public.[231] That which was to be open had become buried in secrecy by some of the same people who limited access to knowledge in the past. The public outcry they sought to avoid is now justifiably being made by many opposed to such secrecy.

In November, 1993, about 400 scientists gathered at John Hopkins University Applied Physics Lab to discuss their work in developing non-lethal weapons technologies, including radio frequency radiation (RF), electromagnetic pulse (EMP), ELF fields, lasers and chemicals. The meeting was classified, and no detailed reports were ever publicly released.[232] According to the press statements and the conference agenda (which was released) the programs developing the technologies had made significant advances. Enough advancement had been made to establish a secrecy veil and classify the conference. This conference took the whole program of nonlethal weapons a step forward by bringing the leading experts together for this

231. "Nonlethal Arms, New Class of Weapons Could Incapacitate Foe Yet Limit Casualties, by Thomas E. Ricks, *The Wall Street Journal,* January 4, 1993, pp A1 and A4.
232. "Military on Nonlethal Weapons: 'A Very Attractive Option'", *Microwave News,* November/December, 1993.

event. The conference was sponsored by Los Alamos National Laboratory and focused on both military and law enforcement uses for these technologies. Dr. Edward Teller and U. S. Attorney General Janet Reno were the scheduled keynote speakers at the conference, although Reno was unable to attend.[233] (The same Dr. Teller was the "father of the H-bomb".)

Dr. Clay Easterly of Oak Ridge National Laboratories led a session on ELF/EMFs. "My major point was that there seem to be some biological sensitivities or responses to ELF fields that could be useful for nonlethal technology." He could not discuss the specific applications because the information was classified by the military. He went on to say that the military was interested in the use of nonionizing (non-thermal) radiation which could be used for disabling enemy electronics, although his presentation dealt with uses which would affect people.[234]

A number of papers were presented at the conference that gave the latest information on applied technologies in this area. Dr. Henry Brisker from the U. S. Army Research Laboratory presented work dealing with High Power Microwave Technology, while Dr. George Baker of the Defense Nuclear Agency presented a paper titled "RF Weapons: A Very Attractive Nonlethal Option".

What is obvious from the number of participants and the scope of the subject matter is that the development of these ideas in new military hardware did not just arrive on the scene, but rather, had been a part of the military agenda for some time. The current emphasis on these technologies reflects the level of interest of the government; and the increased visibility

233. "Military on Nonlethal Weapons: 'A Very Attractive Option'", *Microwave News*, November/December, 1993.
234. Ibid.

the government is giving these areas indicates an intention to use these systems in a more open manner.[235] It is common practice to let things like these types of technological advances come out in small bits in order to "test the waters" of public opinion. In this way the population can be convinced to accept greater levels of intrusion by governmental agencies. The idea is to indoctrinate by being taught to believe, rather than being given all of the facts so that a prudent person could think about the issues and make reasoned decisions. The contrast can be better stated as; propaganda -versus- persuasion by reasoned debate.

These technologies extend beyond the Air Force, CIA and Navy. "*The U. S. Army has its own product - a radio frequency (RF) weapon. In the August 24, 1987 issue of Defense News, John Rosado of Harry Diamond Labs in Adephi, MD, is quoted as saying that 'the nature of warfare will be completely changed by the use of (RF) weapons.' Rosado also noted that RF radiation could be used over a wide battlefield with phased arrays..."*.[236] A phased array antenna system is what HAARP uses to send its signals to the ionosphere. On a smaller scale they can be rapidly deployed anywhere in the world.

From National Defense to The Justice Department

On July 21, 1994, Dr. Christopher Lamb, Director of Policy Planning, issued a draft Department of Defense directive which would establish a policy for non-lethal weapons. The policy was intended to take effect January 1, 1995, and formally connected the military's non-lethal research to civilian law enforcement agencies.

235. *Nonlethal Defense, a Classified Conference Sponsored by Los Alamos National Laboratory*, November 16-17, 1993.
236. *Microwave News*, May/June, 1988.

The government's plan to use pulsed electromagnetic and radio frequency systems as a nonlethal technology for domestic Justice Department use rings the alarm for some observers. Nevertheless, the plan for integrating these systems is moving forward. Coupling these uses with expanded military missions is even more disturbing. This combined mission raises additional constitutional questions for Americans regarding the power of the federal government.[237]

In interviews with members of the Defense Department, the development of this policy was confirmed.[238] In those February, 1995 discussions, it was discovered that these policies were internal to agencies and were not subject to any public review process.

In its current draft form, the policy gives highest priority to development of those technologies most likely to get dual use, i.e. law enforcement and military uses. According to this document, non-lethal weapons are to be used on the government's domestic "adversaries". The definition of "adversary" has been significantly enlarged in the policy:

"The term 'adversary' is used above in its broadest sense, including those who are not declared enemies but who are engaged in activities we wish to stop. This policy does not preclude legally authorized domestic use of the nonlethal weapons by United States military forces in support of law enforcement."[239]

This allows use of the military in actions against the citizens of the country that they are supposed to protect or a declared enemy. This policy statement begs the question; who

237. Department of Defense Directive, *Policy for Nonlethal Weapons*, Office of the Assistant Secretary of Defense, Draft, July 21, 1994.
238. Interviews in late February, 1995 by Nick Begich.
239. Department of Defense Directive, *Policy for Nonlethal Weapons*, Office of the Assistant Secretary of Defense, Draft, July 21, 1994.

are the enemies that are engaged in activities they wish to stop, what are those activities, and who will make the decisions to stop these activities?

An important aspect of non-lethal weapon systems is that the name non-lethal is intentionally misleading. The Policy adds, "It is important that the public understand that just as lethal weapons do not achieve perfect lethality, neither will 'non-lethal' weapons always be capable of precluding fatalities and undesired collateral damage".[240] In other words, you might still destroy property and kill people with the use of these new weapons.

Pentagon officials said they would like to get $41 million in 1995 for developing these weapons that include their electromagnetic pulse systems.[241] This "meager" sum has a good deal of reach, because it is for implementing already understood technology, for building prototype weapons out of mostly "off the shelf" parts. The only challenge would be putting them together. The basic research for these technologies was done with "black budget" money. (These are funds which are so secret even the Congress does not know how the money will be spent, behind which organizations like the CIA, government laboratories run by the military and other intelligence organizations, hide their programs.) The Pentagon actually received $50 million to be used jointly with the Department of Justice in developing these "non-lethal" weapons.[242] Significant funding has been made available every year since 1995, in both DoD, Department of Justice and other budgets of the federal government.

240. Ibid.
241. "Nonlethal Arms, New Class of Weapons Could Incapacitate Foe Yet Limit Casualties", by Thomas E. Ricks, *The Wall Street Journal*, January 4, 1993, pp A1 and A4.
242. "High-Tech Civilian Control Studied; Secret Pentagon - DoJ Memo of Understanding, 'Non-lethal' Weapons Under Development, are Being Added to the Government Arsenal in its War Against its Own Citizens", Warren Hough, *The Spotlight*, July 31, 1995.

In press statements, the government continues to downplay the risks associated with such systems, even though the lethal potential is described in the context of their own usage policy. In Orwellian double speak, what is nonlethal can be lethal.

The development of these technologies is being jointly managed by the Nonlethal Weapons Steering Committee, which is co-chaired by the Undersecretary of Defense for Acquisition and Technology, and the Office of the Assistant Secretary of Defense for Special Operations and Low Intensity Conflict.[243] This crystallizes the new Justice Department and Department of Defense alliance for future law enforcement and military initiatives in the United States.

The weaving together of Department of Defense missions with civilian Department of Justice missions is unprecedented. Not since the Civil War has the military machinery – except in very limited riot control actions – been turned against United States citizens. The idea of using these dangerous and intrusive systems is counter to good public policy. This raises serious constitutional questions regarding use of the Department of Defense for domestic police actions, which may be in conflict with the narrowly-defined federal use of the military "for the national defense".

International Alarm Bells Begin to Sound

Questions are not just being raised by the author of this book, they are being raised by the International Committee of the Red Cross. In their report from mid-1994,[244] a number of points were made.

243. "Perry Plans to Launch Nonlethal Warfare Effort", *Defense News*, September 19-25, 1994.
244. "Expert Meeting on Certain Weapon Systems and on Implement-ation Mechanisms in International Law", Report of the International Committee of the Red Cross, Geneva, Switzerland, May 30 - June 1, 1994. Issued July 1994.

The idea of "war without death" was not new but began in the 1950's, according to the report. The military interest in these systems dealt with chemical weapons, later advancing to radiation weapons. The report looked at the ramifications of international law regarding use of these new technologies. It pointed out weaknesses in the international conventions regarding the use of chemical weapons:

> *'Therefore, when the Convention (Chemical Weapons Convention) comes into force next year, activities involving them - activities such as development, production, stockpiling and use - will become illegal, unless their purpose is a purpose that is expressly not prohibited under the Convention. One such purpose is 'law enforcement including domestic riot control purposes'[245] Unfortunately, the Convention does not define what it means by 'law enforcement' (whose law? what law? enforcement where? by whom?), though it does define what it means by 'riot control agent', namely 'any chemical...which can produce rapidly in humans sensory irritation or disabling physical effects which disappear within a short time following termination of exposure'. States parties are enjoined 'not to use riot control agents as a method of warfare.''"[246]*

In other words, we can use on our own citizens what we cannot use in warfare with real enemies who are threats to national security. This explains why the development of nonlethals has moved out of the Department of Defense into the Department of Justice. For the Department of Defense to continue to work on these weapons, as instruments of war, is now illegal under international law. The Red Cross report went on to discuss the shift from weapons of war to police tools which they called "riot control agents".

245. Chemical Weapons Convention, Article II.9(d).
246. "Expert Meeting on Certain Weapon Systems and on Implementation Mechanisms in International Law", Report of the International Committee of the Red Cross, Geneva, Switzerland, May 30 - June 1, 1994. Issued July, 1994.

What does this mean for Americans? This places Americans, and citizens of other counties, in a lesser protected class than individuals seeking to destroy our countries, our real adversaries. This language really represents a way for countries to continue to develop these weapons. This is a loophole in the agreement. So while the treaty looks good on the surface, it is hollow rhetoric underneath.

In another section of the report, "Future Weapons Using High Power Microwaves" are discussed at length. This section describes microwave frequencies developed for use in weapons against machines and people.

One of the uses described is an Electromagnetic Pulse (EMP) weapon which gives an operator the same ability to wipe out electronic circuits as a nuclear blast would provide. The main difference is that this new technology is controllable, and can be used without violating nuclear weapons treaties.

This section of the report then described energy levels needed for the following to occur:

- "Overheats and damages animal tissue."
- "Possibly affects nervous system."
- "Threshold for microwave hearing."
- "Causes bit errors in unshielded computers."
- "Burns out unprotected receiver diodes in antennas."

The effects are based on radio frequency radiation being pulsed "between 10 and 100 pulses per second". The report confirmed that non-thermal effects were being researched. These non-thermal effects included damage to human health when the effects occurred "within so-called modulation frequency windows or power density windows".[247]

The way these weapons work was clearly described when the report noted their effect on machines:

> *"A HPM (High Power Microwave) weapon employs a high power, rapidly pulsating microwave beam that penetrates electronic components. The pulsing action internally excites the components, rapidly generating intense heat which causes them to fuse or melt, thus destroying the circuit."*

Tesla described his speed of light system as being able to melt aircraft hundreds of miles away. The Red Cross report echoes Tesla:

> *"HPM (weapons) attack at the speed of light thus making avoidance of the beam impossible, consequently negating the advantage of weapon systems such as high velocity tactical missiles."*

In other words, with this kind of weapon there is no machine which could get by this invisible wall of directed energy.

These technologies have been, and will continue to be, increasingly "frequency code" specific and more precisely tuned to create the "controlled effects" desired by military planners. Dr. Carlo Kopp is an expert in Computer Science and Information Warfare and authored a series of papers in the 1990's on e-bombs (electromagnetic bombs). He defined this new bomb with a formal definition as, "any non-nuclear bomb which produces its primary damage effect by the use of

247. "Expert Meeting on Certain Weapon Systems and on Implementation Mechanisms in International Law", Report of the International Committee of the Red Cross, Geneva, Switzerland, May 30 - June 1, 1994. Issued July, 1994.

electromagnetic fields or waves." He goes on to discuss how such bombs can discharge energy that will either enter by **coupling** with the antennas on mobile or wireless devices or through "backdoor **coupling** via network cables, main power wiring and telephone wiring, but also via cooling grilles and air gaps in computer or other electronic equipment chassis."[248] This type of technology is understood in China and Russia as we have previously reported. The advances in these areas are sure to yield published frequency code specific systems that could be used in either a "preemptive war", a terrorist assault, a corporate goal or government action. These technologies exist and will be reverse engineered, or even created, by our ideological and commercial adversaries. These types of weapons can also be developed for use against electronic circuits as Dr. Kopp stated in his interview. However, when we apply the knowledge of the frequency codes of the body, with the lower energy requirements of targeting the human body, a very different set of e-bomb possibilities present themselves. He suggests as a countermeasure to foreign threats that the United States and others develop infrastructures like fiber optics and electromagnetic hardening technologies for use in building all computers and appliances. The points that he also makes is that "The electromagnetic interference in my suburb has gone up by at least 10dBs, due to the vast number of household computers and digital appliances. Adopt a binding and well constructed electromagnetic hardening standard and the problem goes away, killing two birds with one stone." If the technologies are well thought out we could also guard against any incoming signal in conflict with our health or free will.

Computer hackers are engaged in "infowars" which include the concepts of e-bombs. E-bombs will likely take many forms in coming decades. The idea of using any energy

248."The E-bomb Threat and WMD Terrorism, International Security Research & Intelligence Agency, By Dr. Karen Carth for ISRIA, June 28, 2006. EPI6065

emitting source as a platform for an e-bomb that affects our minds or health is not lost on many governments and others. Gaining access to the United States, or the world's infrastructures, that deal with energy transfers via communication, energy distribution, information systems and any other carrier can be used to piggyback other signals, which can be used for creating great harm, or undo influence, on targeted individuals, groups or entire populations. There are now over 100 countries, and countless private persons, corporations and others, that can even hack through the best military system firewalls. The hack through private sector critical infrastructure security systems will be easily accomplished by those desiring to do so. Attempts to hack through the Pentagon's computers has increased from fewer than 800 tries in 1996 to over 160,000 attempts in 2005. Some of these were successful including Chinese spies that hacked into the Joint Chief's classified computer system stealing significant information regarding national defense.[249]

Safeguards need to be created to protect people and national security at the same time. Presently under the existing law there is no clear way to guard against abuse by the designers of these systems so that they are not used incorrectly or illegally. There needs to be serious criminal consequences to those who would violate the ethical use of this technology. There needs to be international oversight if the issue, with protection for whistleblowers from both government and the private sectors.

An attempt was made by the United States House of Representatives in HR1317 and HR5112 to extend whistle-blower protection to employees of all federal agencies. The United States Senate did just the opposite with the passage with a 96-0 vote on S494 which exempts, "specifically,

249. "Hackers Invade Defense Systems", By Siobhan Gorman, *The Baltimore Sun*, Anchorage Daily News, July 3, 2006, pg 1, EPI6075

workers at the FBI, the CIA, Defense Intelligence Agency, National Imagery and Mapping Agency, and the National Security Agency would not receive the broader protections."[250], [251] At this time the final outcome has not been decided, given the history, we will follow the Senate's lead and continue the disaster of this public policy. Accountability and national security can exist together. The issue of meaningful "whistleblower" protection law must be addressed in the United States, and internationally, with the creation of institutions that will protect humanity and the interests of all people.

Another report on non-lethal technologies, issued by the Council on Foreign Relations points out that, "The Nairobi Convention, to which the United States is a signatory, prohibits the broadcast of electronic signals into a sovereign state without its consent in peacetime".[252]

This report opens discussion of the use of these weapons against terrorists and drug traffickers.[253]The CFR report recommends that this be done secretly so that the victims do not know where the attack is from, or if there even is an attack. There is a problem with this approach. The use of these weapons, even against these kinds of individuals, may be in violation of United States law in some cases in that it presumes guilt rather than innocence. In other words the police, CIA, DEA, FBI, Homeland Security or other enforcement organization becomes the judge, jury and executioner.

250. "Senate Approves Whistleblower Rights Breakthrough", Government Accountability Project, Dylan Blaylock, Communications Director, June 23, 2006. EPI6082
251. "Senate Excludes Intelligence Employees from Whistleblower Bill", by Jonathan Marino, *Government Executive Magazine*, June 26, 2006. EPI6083
252. "Nonlethal Technologies; Military Options and Implications" *Report of an Independent Task Force sponsored by the Council on Foreign Relations*, Malcom H. Weiner, Chairman, released June 22, 1995.
253. Ibid.

New Toys for Defense

In a summary document produced at Maxwell Air Force Base, the real potentials of these weapon applications are described. More important politically is the "Newt Gingrich spin" placed on the development efforts in the foreword of the military's report on this subject.

The foreword to *Low-intensity Conflict and Modern Technology* was written by Congressman Newt Gingrich in 1986, before he rose to the position of Speaker, a position that made him one of the most powerful people in the United States government. Gingrich's views on "low intensity conflicts" become relevant when considered along with the development of non-lethal technology. He focuses on creating approaches to conflict which can dispense with radicals who, in his words, "engage quietly in dirty little wars in faraway places with almost no regard for legal nicety or the technical problems of international law".[254]

The ability of the United States to meet these challenges is also discussed in the foreword as Gingrich wrote:

> *"The organization of power in the State and Defense Departments and the relationships between Congress, the news media, and the executive branch are all unsuited to fighting low-intensity conflict effectively."*

He goes on to describe the limits of the military in addressing these types of conflicts, and commends the military for developing the policy doctrine and the new technology directions of the Air Force put forward in the book. Newt

254. *Low-Intensity Conflict and Modern Technology*, Lt. Col. David J. Dean, USAF Editor, Air University Press, Center for Aerospace Doctrine, Research, and Education, Maxwell Air Force Base, Alabama, June, 1986.

Gingrich apparently remains a fan of non-lethal weapons, which he views as useful technology for domestic law enforcement. Specifically he said, when commenting on these technologies, that they "are our real peace dividend" and that they will "preserve the defense industrial base, stimulate jobs in high-technology industry, and provide needed new options to local police and law enforcement authorities."[255]

Another interesting issue to contemplate is the Defense Department's new Policy on non-lethal weapon systems, where "adversaries" and "enemies" have been more broadly defined here as the use of such systems against American citizens. The idea that these systems can be used almost without detection to manipulate the behavior and thinking of people raises moral questions. This use also appears to be in conflict with Constitutional rights regarding free expression and speech. The fact that the military together with the United States Justice Department can now use these technologies, under their broad definitions, should sound an alarm for all people, not just Americans.[256]

The only redeeming discussion in *Low-Intensity Conflict and Modern Technology*, (a compilation of papers presented in 1984) was that some at the meeting saw the "new" technologies for what they were, and began heated debates about the morality of using these weapon systems. In the closing page of the technology overview the following was written:

> *"Paul Tyler also discusses the application of electromagnetic radiation (EMR) to low-intensity conflict. He surveys ongoing scientific research into the biological effects of electromagnetic radiation. Tyler*

255. "Armageddon Killing Them Softly", by Russell Shorto, GQ, March, 1995.
256. U.S. Department of Defense Directive, *Policy for Nonlethal Weapons*, Office of the Assistant Secretary of Defense, Draft, July 21, 1994.

tells us current evidence indicates that specific biological effects can be achieved by controlling the perimeters of electromagnetic radiation directed at human subjects. Thus there is the potential to use EMR to control human behavior or even to maim or kill adversaries. Tyler urges that the United States should devote considerable resources to exploring the possibilities of developing EMR weapons technology, which could be of particular value in low-intensity scenarios."[257]

It becomes increasingly obvious that the inner circle of the Department of Defense placed a substantial amount of effort and emphasis in these areas.

The writer continued:

"Both Tyler's and Ruotanen's (another contributor to 'Low-Intensity Conflict and Modern Technology' papers created heated discussions. Some panel members questioned the advisability of employing nuclear, EMP (electromagnetic pulse), and EMR (electromagnetic radiation) weapons. They felt the ever-present danger of escalation would negate any advantage to be gained from surgical ground or atmospheric nuclear burst. Some on the panel saw Tyler's article as bordering on moral heresy. It is acceptable to have weapons and strategies to blow bodies into little pieces or burn them to a crisp, but not to use medical research and techniques to develop more subtle ways of eliminating or controlling enemies. There was some feeling that any benefit to be gained from research into EMR effects - dangerous in itself - might well be overshadowed by indignant outcries against the use of EMR weapons on human beings."[258]

257. *Low-Intensity Conflict and Modern Technology*, Lt. Col. David J. Dean, USAF Editor, Air University Press, Center for Aerospace Doctrine, Research, and Education, Maxwell Air Force Base, Alabama, June, 1986.
258. Ibid.

The development of these new weapons helps demonstrate the general immorality of war, particularly when the military establishment presents mind manipulation as a preferable outcome to death.

Another revealing section of this military compiled book is apparently aimed at a rationale which would make sense in the minds of defense planners and would help them deal with the moral issues:

> *"The articles and discussions in the technology panels generated quite a bit of heat and quite a bit of light. Chief among the concepts brought into the light was that the application of technology to low-intensity conflict should not hinge on the debate over simple or advanced systems, or high versus low technologies...The key is to recognize specific requirements and apply suitable systems and technologies to meet those requirements."*[259]

Stated differently, the military's means justifies the military's end.

Obviously, these concepts were well understood by the military in 1984 when the original work was presented which led to the above referenced book's publication in 1986. The system's development moved forward – most likely within "black budgets" – It was *1984*.

The next chapter in *Low-Intensity Conflict and Modern Technology* is the section written by Captain Paul Tyler. Tyler discusses, to some degree, the application of non-ionizing radiation using external fields, including radio frequency radiation and other electromagnetic radiations. He discusses

259. *Low-Intensity Conflict and Modern Technology*, Lt. Col. David J. Dean, USAF Editor, Air University Press, Center for Aerospace Doctrine, Research, and Education, Maxwell Air Force Base, Alabama, June, 1986.

some of the beneficial effects of this energy for healing wounds, bone regeneration in fractures, electroanesthesia, acupuncture and pain relief, all based on the frequency codes carried by the energy source.

He then jumps into the military applications of this technology:

> *"The potential applications of artificial electromagnetic fields are wide-ranging and can be used in many military or quasi-military situations...Some of these potential uses include dealing with terrorist groups, crowd control, controlling breaches of security at military installations, and antipersonnel techniques in tactical warfare. In all of these cases the EM systems would be used to produce mild to severe physiological disruption or perceptual distortion or disorientation. In addition, the ability of individuals to function could be degraded to such a point that they would be combat ineffective. Another advantage of electromagnetic systems is that they can provide coverage over large areas with a single system. They are silent and countermeasures to them may be difficult to develop... One last area where electromagnetic radiation may prove of some value is in enhancing abilities of individuals for anomalous phenomena."*[260]

Tyler's comments point to applications which may already be somewhat developed. He refers to an earlier Air Force document about the uses of radio frequency radiation in combat situations. He also points out that the uses may include enhancement of "anomalous phenomena" in individuals. Anomalous phenomena are those kinds of things

260. *Low-Intensity Conflict and Modern Technology*, Lt. Col. David J. Dean, USAF Editor, Air University Press, Center for Aerospace Doctrine, Research, and Education, Maxwell Air Force Base, Alabama, June, 1986, pp 249-251.

which are not readily explained. You might say these are the quirks in human potentials. These could be called extrasensory or supernormal phenomena. This significant area is carefully screened in the writing. He refers to Eastern European research and the Soviet work also, without disclosing what exactly they were doing with this technology. There are a number of records on this in the research.[261, 262, 263]

Another document prepared for the Air Force in 1982, reveals the direction that branch of the armed services took the technology. While commenting on radio frequency radiation uses the report stated:

> *"Biotechnology research must consider the significant advances that can be made in electromagnetic radiation weapons and defenses that could be in place by the year 2000...Research is first needed to develop and apply methods for assessing pulsed RFR (radio frequency radiation) effects. Techniques are needed for depositing RFR at selected organ sites. Mathematical models and physical measurement capabilities must be developed to track, real time, RFR energy distributions within these organ sites as a function of physiological processes such as diffusion and blood flow. These studies will require prudent extrapolation of physical and physiological data obtained from laboratory animals to humans in operational environments."*[264]

261. Defense Intelligence Agency, DST-1801S-387-75, *Soviet and Czechoslovakian Parapsychology Research (U)*, Prepared by U.S. Army Medical Intelligence and Information Agency, Office of the Surgeon General, Sept., 1975, pp 1-71 EPI6060
262. Defense Intelligence Agency, ST-CS-01-169-72, *Controlled Offensive Behavior – USSR (U)*, Prepared by the US Army Office of The Surgeon General Medical Intelligence Office,Author Captain John D. LaMothe, July, 1972, pp 1-173 EPI6061
263. Defense Intelligence Agency, DST-1810S-202-78, *Paraphysics R&D - Wasaw Pact (U)*, prepared by U.S. Air Force , Air Force Systems Command Foreign Technology Division, Authors: *classified*, March, 1978, pp 1-130, EPI6062
264. *Final Report On Biotechnology Research Requirements For Aeronautical Systems Through the Year 2000*, Volume I, Southwest Research Institute, San Antonio, Texas, 1982, pp 44.

This 1982 report called for research to develop hardware that could deliver the radiation through new weapons for the Air Force.

A research publication commissioned by the Air Force – *Radio Frequency Dosimetry Handbook* – described these models, in time to meet the schedules they had put forward in 1982.[265] The book gives mathematics for calculating the dosages (Dosimetry) of radio frequency radiation necessary to cause changes in animals and humans. The book compiled research spanning five decades, and contains a bibliography which is 29 pages long. In 1996 it was the only book of its kind available on the subject in the world. This book, and more importantly the research behind it, forms the matrix of thought and theory for the finest electromedical devices for healing yet contemplated. However, this military research is not being used to heal; it is being used to develop better methods of killing.

One of the principal reviewers of the *Radiofrequency Radiation Dosimetry Handbook* for the Air Force was Dr. Herman P. Schwan. Dr. Schwan was also the largest contributor to the book, having two full pages of his work alone cited in the bibliography. His basic research is being used for the development of military hardware for use by the West. His employment in this effort is another case of the "military means justifying the military ends". An interesting aside to this individual's history includes his work in Nazi Germany during World War II, where he worked in scientific laboratories. He was later admitted to the United States under a special military operation called "Project Paperclip"[266], a program designed to bring to the United States prominent scientists who had been

265. *Radiofrequency Radiation Dosimetry Handbook*, United States Air Force School of Aerospace Medicine, Brooks Air Force Base, Texas, October, 1986.
266. *Project Paperclip*, by Dr. Clarence Rasby, Athenaeum Press,1975

employed in Nazi laboratories during the war.[267] In 1957, Dr. Schwan was working for the Biological Warfare Laboratories at Fort Detrick, Maryland, where the United States military gave him a "Top Secret" security clearance.[268] Dr. Schwan was working at University of Pennsylvania in 1995.

Going back for a moment to the 1982 Air Force report referred to earlier, the use of radio frequency radiation in the new weapon systems contemplated by the Air Force was intended to change the central nervous system, cardiovascular system and the respiratory system. I was distressed to read about the idea of using this technology to alter the way people think in order to make the "enemy" of the government incapable of waging war.[269] This type of use of a technological advancement is inherently evil. Thinking is a fundamental right of all people. No individual or government should interfere with the free will of individuals in this way. Moreover, the fact that these technologies are classified – and hidden from the medical community – denies humanity the opportunity to explore these areas for healing people rather than for destructive purposes.

The report speaks about using this technology in a way which could interact with biological or chemical agents. The report states, "it may be possible to sensitize large military groups extremely dispersed amounts of biological and chemical agents. It should be noted that this may require relatively low

267. Memorandum, Office of the Deputy Director of Intelligence, Headquarters European Command, October 30, 1947. (Provided by: International Committee for the Convention Against Offensive Microwave Weapons)
268. Memorandum, Headquarters Second United States Army, Fort Meade, Maryland, September 5, 1957. (Provided by: International Committee for the Convention Against Offensive Microwave Weapons)
269. *Final Report On Biotechnology Research Requirements for Aeronautical Systems Through the Year 2000,* Volumes I and II, Southwest Research Institute, San Antonio, Texas, 1982.

level RFR."[270] The idea expressed here is that depositing small amounts of chemicals in a person's body, in amounts below normal levels where negative physical effects are known to occur, will ensure they have no perceivable effect until radio frequency radiation (RFR) is introduced. Once introduced, the RFR creates physiological reactions which are detrimental to the individual host. This would allow individuals who are not exposed to the chemicals to then enter the area of the RFR without harm to their own bodies. When an operator tunes the RFR in just the right way, changes are caused in the energy state of atoms, which cause chemical reactions in the body, which in turn manifest as physiological or psychological changes, all based on the frequency codes.

The introduction of small amounts of chemicals is a very important concept when cyclotron resonance is considered. Cyclotron resonance occurs when, as a result of complex electromagnetic interactions, a charged particle or ion begins to go into a circular, or orbital, motion. The speed of the orbits is determined by the ratio between the charge and weight of the particle and the strength of the applied field. Once this happens, and a specific charged particle frequency code determined, if an electric field is added that is in harmony with this frequency at right angles to the magnetic field, energy is transferred from the electric field to the charged particle.

Many activities in living cells involve charged particles. Cyclotron resonance allows for the transfer of energy to cause ions to move more rapidly. It is cyclotron resonance which allows very low strength electromagnetic fields, together with Earth's magnetic field, to produce major biological effects. It happens because the total effect impacts very specific particles when tuned to the right frequency codes. To some degree, this

270. *Final Report On Biotechnology Research Requirements For Aeronautical Systems Through the Year 2000*, Volumes I and II, Southwest Research Institute, San Antonio, Texas, 1982.

explains why nonionizing levels of radiation produce the effects which they do.[271] The right weapon system could literally be tuned like a radio by an operator for the maximum negative effect. The principle could also be applied for healing purposes, and this area is being researched in some quarters.

When combined with the Earth's normal magnetic fields, it is important to note, ELF (1-100 Hertz) frequencies can be carried on any pulsed carrier which in some window frequencies can cause controlled or accidental negative health effects.

The concept of cyclotron resonance was applied to the research carried out by the U.S. Naval Medical Research Center.[272] They affected a chemical which was already in the body, or one that is introduced to the body, a person could literally amplify the effect of the chemical to the point where a very significant effect is caused, even death. Many naturally occurring chemicals are not dangerous in the amounts present in the body. If these chemicals are affected by RFR, however, their potency increases to the point of being dangerous or, if used for positive effects, correcting imbalances in the body, sparing the liver and kidneys from the impact of pharma-based approaches to health. The introduction of RFR causes a change in energy states, which causes chemical reactions. The results depend on wave form, frequency and energy level introduced, among other things.

The 1982 report, Final Report on Biotechnology Research Requirements for Aeronautical Systems Through the Year 2000, went on to describe graphically the research efforts, breaking them into three primary areas.

271. *Cross Currents, The Perils of Electropollution, The Promise of Electromedicine*, by Robert O. Becker, M.D., pp 236-237.
272. *Cross Currents, The Perils of Electropollution, The Promise of Electromedicine*, by Robert O. Becker, M.D., pp 236-240.

• "Pulsed RFR Effects" which covered a research period from 1980 through 1995, emphasized "considerably increased efforts" in this area.

• "Mechanisms of RFR with Living Systems" was referred to as a "continuation of ongoing research" beginning in 1980, and forecast to conclude around 1997.

• The last area of emphasis was "RFR Forced Disruptive Phenomena", which was considered a "major new initiative". It was to begin around 1986 and continue until 2010. This last area could be characterized as applied technology, scheduled to begin at the same time as the *Radiofrequency Radiation Dosimetry Handbook's* completion. The handbook gave the basic information needed to develop new and powerful weapon systems which would have a negative effect on humans and electronics, but would leave other property unscathed.[273] By the mid-1990's, the timeline for development was on or ahead of schedule.

The above report grew into a second volume, which gave a more detailed account of the Air Force's 1982 level of knowledge of radio frequency radiation (RFR) impacts on people. This report says, "As the technological race continues, knowledge of mechanisms of action of RFR with living systems and the assessment of pulsed RFR effects will demonstrate the vulnerability of humans to complex pulsed electromagnetic radiation fields in combination with other stresses."[274]

273. *Final Report on Biotechnology Research Requirements for Aeronautical Systems Through the Year 2000*, Volumes I and II, Southwest Research Institute, San Antonio, Texas.
274. *Final Report on Biotechnology Research Requirements for Aeronautical Systems Through the Year 2000*, Volumes I and II, Southwest Research Institute, San Antonio, Texas, pp 181-188.

The Air Force document laid out the aims and approaches for developing technological expertise to make adversaries helpless. The use of the technologies crystallized in the *Radiofrequency Radiation Dosimetry Handbook.* The researchers' intent was to monitor the effects of the radio frequency radiations at a chemical, molecular and atomic level. The document stated:

> *"Currently available data allow the projection that specially generated radio frequency radiation (RFR) fields may pose powerful and revolutionary antipersonnel military threats. Electroshock therapy indicates the ability of induced electric current to completely interrupt mental functioning for short periods of time, to obtain cognition for longer periods and to restructure emotional responses over prolonged intervals. Experience with electroshock therapy, RFR experiments and the increasing understanding of the brain as an electrically mediated organ suggest the serious probability that impressed electromagnetic fields can be disruptive of purposeful behavior and may be capable of directing and/or interrogating such behavior. Further, the passage of approximately 100 milliamperes through the myocardium can lead to cardiac standstill and death, again pointing to a speed-of-light weapons effect."*[275]

The advances in wireless transmission of energy for these kinds of weapons has reached the stage where we could see them used in the coming decade. It is likely that the research will give technicians increasingly more versatile and controlled effects.

275. *Final Report On Biotechnology Research Requirements For Aeronautical Systems Through the Year 2000*, Volumes I and II, Southwest Research Institute, San Antonio, Texas, pp 181-188.

The document then built into a wild scheme which, upon first reading, escaped our research focus. We couldn't quite get the idea of what the Air Force meant by "interrogating", and how this would fit into the picture of radio frequency radiation weapons hardware.

It became clear when we read the following:

"While initial attention should be toward degradation of human performance through thermal loading and electromagnetic field effects, subsequent work should address the possibilities of directing and interrogating mental functioning, using externally applied fields within the possibility of a revolutionary capability to defend against hostile actions, and to collect intelligence data prior to conflict onset."[276]

What this seemed to say was that the objective of the research would be toward mind manipulation at a distance, where the military could alter what people thought and, at the same time, know what they thought. It seemed too much like a science fiction novel. How in the world could this effect be possible? How long had research been underway, so that the military would be so bold as to publish this kind of material in an unclassified document? Could it be that they had quietly advanced their research into these technologies under a veil of secrecy? Then again, the technology had been around since the 1960's. It had been thoroughly investigated and most likely used in forming the basis for new weapons.

The report went on to describe a system which again pointed to a radio frequency radiation (RFR) project. The report called for the development of:

"A rapidly scanning RFR system [which] could provide an effective stun or kill capability over a large area. System effectiveness will be a function of

276. Ibid.

waveform, field intensity, pulse widths, repetition frequency and carrier frequency. The system can be developed using tissue and whole animal experimental studies, coupled with mechanisms and waveform effects research."

The grossly unfortunate part of this type of experimentation is how it is used. Using RFR and other energies to destroy human life, as opposed to enhancing it, is immoral. The new directions of the energy sciences in healing people and enhancing health are not considered by the military – a travesty and a disservice to us all. At a time when health care costs are rising, and the threats of major conflict declining, our government continues to spend billions on killing rather than on enhancing the human experience. These areas are important, and the research should be available so that more prudent uses can be developed. If this happened, I believe it would speed the positive developments being pursued by those who have great energy and creativity, but limited resources. These methods of healing will replace current pharmacological models for most mental and physical disorders. Using the frequency codes healing will be significantly changed and increased in efficiency.

Unfortunately, the idea of using RFR for destructive purposes negates the entire area of positive life science applications. The research in life sciences using electromagnetic radiations (EMR) is advancing outside of the mainstream medical establishment, but is beginning to be used there also. Some genetic engineering researchers for example, are already using this knowledge in their work.

Going back to the text by Captain Paul Tyler, we can look at the debate between classical theories and recent research. There is a gulf of conflict between these two schools of thought. The debate centers on the classical idea that only ionizing radiation (that which generates heat in tissue) can cause reactions in the body, while new research indicates that subtle, small amounts of energy can cause reactions as well. What Tyler wrote in 1984, as an officer in the Air Force, puts the debate simply. He said,

> *"Even though the body is basically an electrochemical system, modern science has almost exclusively studied the chemical aspects of the body and to date has largely neglected the electrical aspects. However, over the past decade researchers have devised many mathematical models to approximate the internal fields in animals and humans. Some of the later models have shown general agreement with experimental measurements made with the phantom models and animals. Presently most scientists in the field use the concept of specific absorption rate for determining the Dosimetry (dosages) of electromagnetic radiation. Specific absorption rate is the intensity of the internal electric field or quantity of energy absorbed... However, the use of these classical concepts of electrodynamics does not explain some experimental results and clinical findings. For example, according to classical physics, the frequency of visible light would indicate that it is reflected or totally absorbed within the first few millimeters of tissue and thus no light should pass through significant amounts of tissue. But it does. Also, classical theory indicates that the body should be completely invisible to extremely low frequencies of light where a single wave length is thousands of miles long. However, visible light has been used in clinical medicine to transilluminate various body tissues."*[277]

In other words, the classical theories are partially wrong in that they do not fully explain all of the reactions which are observed in the body. The Navy has abstracted over a

277. *Low-Intensity Conflict and Modern Technology*, Lt. Col. David J. Dean, USAF Editor, Air University Press, Center for Aerospace Doctrine, Research, and Education, Maxwell Air Force Base, Alabama, June, 1986.

thousand international professional papers by private and government scientists which explore these issues. The Navy is well aware of the current research which shows the nonthermal effects of various electromagnetic radiations when they interact with living tissue.

Tyler continues, "*A second area where classical theory fails to provide an adequate explanation for observed effects is in the clinical use of extremely low frequency (ELF) electromagnetic fields. Researchers have found that pulsed external magnetic fields at frequencies below 100 Hertz (pulses/cycles per second) will stimulate the healing of nonunion fractures, congenital pseudarthroses, and failed arthroses. The effects of these pulsed magnetic fields have been extremely impressive, and their use in orthopedic conditions has been approved by the Food and Drug Administration.*"[278]

Even the United States FDA, one of the most vigorous regulatory authorities in the world, accepts these nonthermal effects.

Tyler adds, "*Recently, pulsed electromagnetic fields have been reported to induce cellular transcription (this has to do with the duplication or copying of information from DNA, a process important to life). At the other end of the non ionizing spectrum, research reports are also showing biological effects that are not predicted in classical theories. For example, Kremer and others have published several papers showing that low intensity millimeter waves produce biological effects. They have also shown that not only are the effects seen at very low power, but they are also frequency-specific.*"

278. Ibid.

Tyler goes on to discuss the results of this new thinking and the possible effects of these low energy radiations in terms of information transfer and storage, and their effects on the nervous system. Research has shown that very specific frequencies cause very specific reactions, and once a critical threshold is passed, negative reactions occur.[279] The key to Pandora's box has been turned, and the box has now been opened. Will it result in life or death?

Tyler's paper created a good deal of controversy but was based upon good research and an in-depth knowledge of the subject matter covered. In the decade since its publication, the public debate on the material has continued to grow, even though many of the sources cited in these pages have limited distribution.

The topics in this section could elevate the debate on military technologies. What will these areas of research yield? Should citizens take a greater interest in public affairs and remove the layers of secrecy which mask these research efforts? These are the issues we all face now, and will continue to face as we move further down the road of new technology.

279. Ibid.

Chapter Ten

Who are the Victims?

There are many victims of this technology as a result of the various attempts made by the CIA and others in controlling the human mind. This is a fact based on their own admission and the admissions of other parts of the United States government. Unlike other things hidden from the public because of "national security" making this matter public serves a higher security – *It is a matter of "human security".*

The very essence of what it means to be human is attacked by the misuse of these technologies. Consider that "free will", the right to self-determination, is what separates us as individually created beings and also liberates us in the fullest sense of true liberty. The concept of "private thinking" is fundamental to the right of free speech, assembly, religion and every other God-given right we share in common as human beings. Those rights first require the right to *freedom of thought and mind*. Those that move against the mind violate the human spirit in a manner that most religions of the world agree that God will not even do. Who gives anyone the right to interfere with the "free will" or the mind of another person?

Sometimes referred to as "wavies" or "beamers", victims are usually dismissed when asserting that they are the victims of mind control weapons testing. In fact, at the "University of South Florida researchers have published a study showing that fears of the Internet are replacing the CIA and radio waves as a frequent delusion in psychiatric patients. In every case of Internet delusion documented by the researchers, the patient actually had little experience with computers."[280] The problem is that it is difficult if not impossible to sort out which people might be victims and which are delusional. Attempts to determine the reality of the complaints are often the butt of jokes and fear. For example the "University of Albany has shut down the research of a psychology professor probing the 'X-Files' world of government surveillance and mind control. At conferences, in papers and research over two semesters, Professor Kathryn Kelley explored the claims of those who say they were surgically implanted with communications devices to read their thoughts."[281]

Since the release of my first coauthored book (with Jeane Manning), *Angels Don't Play This HAARP: Advances in Tesla Technology,* I have heard from hundreds of people making such claims. We cannot sort out who might be real victims from those who are not. I believe that the claims should be taken seriously and that people should have someplace to go in order to find the truth or to gain the medical treatments they otherwise deserve. The history of the United States is littered with examples of people being exploited by scientists working under the cover of darkness funded by "black budgets" – could these new victim reports have a factual basis? I believe that some do. I believe that full disclosure and cessation of the misuse of these technologies should be initiated through domestic and international law, in all countries of the world. Freedom of thought and mind should be seen as a basic

280 "Internet Feeds Delusions", Associated Press, July 5, 1999. EPI123
281 "Albany Suspends implants research" Andrew Brownstein, *Times Union,* Aug. 25, 1999. EPI833

fundamental human right. The right to think, free of any form of technical manipulation, must be pressed into the institutions of law and civility.

Government control of the mind in order to impose its will on people is best summarized on the wall of the Franklin D. Roosevelt Memorial, where an inscription appears that reads, "They (who) seek to establish systems of government based on the regimentation of all human beings by a handful of individual rulers...call this a new order. It is not new, and it is not order."[282]

There are things that need to be done in support of victims in the areas covered in this book and elsewhere. The United States must be held accountable for what it has done to its own citizens and countless people outside of U.S. boundaries. At the present time there are U.S. Supreme Court cases that support the suppression of even the names and institutions involved in this kind of classified research giving the government the right to use the finest minds in perpetrating their crimes against humanity. Is this right? Is this the United States of America, the country that our fathers fought to preserve, engaged in this nasty work? Our country is about freedom, liberty and justice, none of which can be realized unless our shared values are amplified in our public policy as technology evolves.

A number of things that can be done by those that support the cause of victims. My present role is public education on these issues. Getting these issues addressed will take refining the data and presenting the evidence as many have done in the past. This time people need to help to do this. Individuals throughout the world send me research reports, articles, papers and bits of information all the time that I use in

282. "Human Research Subject Protection", Anthony N. Lalli, http://www.pw2.netcom.com~allalli/BillSite_analysis/paper_web.html EPI619

lectures, radio reports and through my published work in these areas. Currently, the best materials, on a priority basis, are being added to the database at www.layinstitute.org. Making information available is important for getting the issues addressed. Those who believe they are victims are urged to write to us explaining their situation so that a record of those situations can be kept in the event solutions are found. As solutions develop or ways that people can help address these and related issues, there will be a way to reach individuals and provide information on these subjects.

Consider helping in the effort in support not just of victims, but for everyone on the planet. Whose "will" shall end our "free will"?

Chapter Eleven

Moving Down the Path

The idea that the brain can be made to function at a more efficient and directed level has been the subject of research by scientists, mystics, health practitioners and others for as long as mankind has contemplated such matters. In the last decade, advances in the science of the brain have begun to yield significant research results that are startling, challenging and will be frightening if misused. The only certainty to be expected from the research is that it will continue to proceed.

Where the military would like it to go is of great concern when considered completely. The following is an example:

> **"It would also appear possible to create high fidelity speech in the human body, raising the possibility of covert suggestion and psychological direction...Thus, it may be possible to 'talk' to selected adversaries in a fashion that would be most disturbing to them."**[283]

The concept that people can be impacted by external signal generators which create, for example, pulsed

283 *New World Vistas: Air and Space Power for the 21st Century - Ancillary Volume*; Scientific Advisory Board (Air Force), Washington, D.C.; Document #19960618040; 1996; pp. 89-90. EPI402

electromagnetic fields, pulsed light and pulsed sound signals is not new. The following information demonstrates some of the possibilities and gives hints of the potentials for the technology. On the positive side, researchers in the field of light and sound are making huge progress in a number of areas, including working with learning disabilities, attention deficit disorders, stroke recovery, accelerated learning, drug/alcohol addiction and enhanced human performance. The research has shown that certain brain states can be influenced in a way which causes changes within the brain itself. These changes allow individuals the possibility of influencing specific conditions in the mind and body otherwise thought beyond our direct control.

The military and others interested in such things have also focused a large amount of research into this area for the purpose of enhancing the performance of soldiers while degrading the performance of adversaries.

What is known is that great strides in the area of behavioral control are now possible with systems developed and under development by the most sophisticated countries on the planet. These new technologies represent a much different approach to warfare that our government is describing as part of the *Revolution in Military Affairs*. While these new technologies offer much for military planners they offer even more to citizens in general. Their potential use in military applications and "peacekeeping" creates the need for open debate of this new realm of intelligence gathering, manipulation and warfare. The most basic ethical questions regarding use of these technologies have not been adequately addressed.

At the same time that defense and intelligence gathering capabilities are being sought, independent researchers are fully engaged in seeking positive uses for the technology. The potentials of the science, like all technology, are great as both destructive and constructive forces for change. The idea of enhancing physical and mental performance while bypassing what heretofore was a long and arduous road to achieve the same results is exciting. Maintaining the research in the open literature and ensuring that constructive uses are encouraged is critical.

I began looking into technologies for stimulating brain performance about twenty-five years ago. At the time there were limited tools available compared to what is now possible. Now it is possible to obtain light and sound, electrocranial and biofeedback tools for use in this quest for personal improvement. Moreover, there are audio materials also available for use with most of these tools. These audio materials can be used for learning languages, behavior modification or enhanced performance. The biofeedback side of the new technology is being used to train people to reach specific desired brain states for optimum performance. These tools and devices are discussed in Part II of this book.

The use of light and sound devices for stimulating brain activity for accelerated learning and relaxation, is a growing area of interest to many people. Also, the use of these instruments in conjunction with biofeedback has been the subject of quickly evolving research. The combined technologies of brain state inducement and biofeedback offer exciting possibilities. It has been found, with the combination, a person, in a matter of several weeks, can learn to purposefully modify his/her brain activity in a way which would have taken a Zen master twenty years to accomplish. It has been shown that some children with attention deficit disorders can be taught to regulate their brain activities so that they can learn efficiently without chemicals. Further, it has been shown that recovering stroke and other brain injury victims can recover more rapidly when working with brain biofeedback practitioners and these new tools.

The research is confirming a good deal about human suggestibility and how to influence individual and group behavior. The underlying message that comes with the new technology is the necessity of providing safeguards against misuse. Additionally, recognition of the everyday stimulation we all get and the effect of these information inputs on our learning processes becomes more clear. Human suggestibility, particularly when in a fatigued condition, has been exploited by terrorists, cults, governments and others in pursuit of their own aims. The passive suggestibility of radio and television as we weave in and out of the semi-sleep states is, for the most part, not even recognized. The passive learning situations become even more relevant when we consider how we "receive the news" in our daily lives. The ability to influence thinking, behavior and performance is indeed a two-edged sword.

It is the case with any system that can be *pulse-modulated* to resonate at the frequency codes of the body, including radio, TV, power grids, computer networks, all wireless systems, the earth's magnetic fields and any other system that will allow energy to be transferred or propagated through it can be used to carry information the brain and body will understand and react to. This is the center of the issue. As Dr. Reijo Mäkelä used to say, "it is all about resonance", which represents the corresponding harmonies between energy transmitter and energy receiver. It is like dialing up a station on a radio, only when the transmitter and receiver are in resonance can a person hear the radio station. Such is the case with all components of the human body, organs, cells, molecules, atoms and so on down to the essence of who each person is on an energetic level, from which creation projects us into physical reality. There are profound implications to the manipulation of people on an energetic level through these new technologies.

The 1980's and 1990's were focused on building up the physical body. This century will see a focus on building the mind and optimizing mental performance. The idea of merging the new technologies into education is interesting and also calls into question who will decide what is learned. In the interim, the possibilities are incredible for those interested in such pursuits. The control of our mental function is no different than the control of the muscles in our bodies. Learning to control or coordinate the activity of our minds will propel our bodies through a much more productive and fuller life. The new tools may offer just such opportunities.

If we begin to think about things like a direct download of information into the brain for learning subjects like mathematics and languages these new technologies will be very useful. These kinds of subjects do not require that *value judgments or belief systems* be applied to the information being retained and learned.

Using a direct download of information would be a fantastic way to learn certain subjects. Other subjects would not be suggested using these methods because these other study areas are not formula driven. These subjects require *value judgments that conform to existing belief systems* being applied to reasoning. These new technologies, in some cases, bypass the conscious mind and deliver the information so it is accepted without question by the subconscious and acted upon. This is a very important point when we consider how much value-filled information comes to us in public education systems. State controlled curriculums and compulsory education present real problems because these systems design programs that they "believe" represent the state of knowledge and activity of culture. The question is: who will control the input for our children? Do we even have the right to use these technologies before we are "adults" and able to make our own value judgments based on what our conscious minds create? We each must be in total control of our inputs, experiences, and our personal belief systems, free of any interference. This is called liberty.

These tools will continue to evolve. The first generation systems are referenced in Chapter Thirteen, Tools for the 21st Century. From my own experience over twenty-five years, I have found these systems to be effective and useful in helping reach my higher potentials. There are other new technologies emerging that go beyond the present state of technology. Perhaps one day even the anomalous human potentials will be unleashed. Are people ready for this? The day is coming, and some of us believe that it is already here.

On the other side of the issue is the potential for misuse and exploitation of the science. Military planners, law enforcement officials and others are now seeking the covert use of these technologies for controlling the ultimate "information processor" – The individual person.

Liberty begins as a concept in the mind. Freedom combines with liberty to form the foundation of what we have created as the common values in a democratic republic based on the notion that all people are endowed by their Creator with certain rights. I had not read the below declaration in too long a time...

> "We hold these truths to be self-evident, that all men are created equal, that they are endowed by their Creator with certain unalienable Rights, that among these are Life, Liberty and the pursuit of Happiness. – That to secure these rights, Governments are instituted among Men, deriving their just powers from the consent of the governed. – That whenever any form of Government becomes destructive of these ends, it is the Right of the

People to alter or to abolish it, and to institute new Government, laying its foundation on such principles and organizing its powers in such form, as to them shall seem most likely to effect their Safety and Happiness. Prudence, indeed, will dictate that Governments long established should not be changed for light and transient causes; and accordingly all experience hath shewn, that mankind are more disposed to suffer, while evils are sufferable, than to right themselves by abolishing the forms to which they are accustomed. But when a long train of abuses and usurpations, pursuing invariably the same Object evinces a design to reduce them under absolute Despotism, it is their right, it is their duty, to throw off such Government, and to provide new Guards for their future security."[284]

I went on to read the rest of the document which highlighted the grievances of the American people under King George III of England. It is interesting to read again because many of those grievances have manifested under our own governments today. It is a time to renew the spirit of freedom we once shared in a way which builds our twenty-first century institutions of government to improve the human condition. It is a time to throw off the tyranny of apathy and become active in forming a better democratic republic, a republic that can again become the leader in human rights based on those long-ago made declarations.

284. The Declaration of Independence of the Thirteen Colonies, In Congress, July 4, 1776, The Unanimous Declaration of the Thirteen United States of America.

Part II

Increasing Our Human Potentials

Chapter Twelve

The Brain Drivers

New technologies have incredible possibilities for improving human brain potential. These technologies are being used to improve learning, memory, and for behavior modification. The high-tech tools of the future are here now, and are being introduced into the market place. A leading writer in the area of brain technologies has been Michael Hutchison, who opened this field to everyday people.

As Hutchison describes it, the brain operates within a relatively narrow band of predominant frequencies. There are four basic groups of brain wave frequencies which are associated with most mental activity. The first, beta waves associated with normal activity with the high end of this range associated with stress. The second group, alpha waves can indicate relaxation are ideal for learning and focused mental activity. The third, theta waves indicate mental imagery, access to memories and internal focus. The last, delta waves are found when a person is in deep sleep.[285]

External stimulation of the brain by electromagnetic means can cause the brain to be entrained or locked into phase with an external signal generator. Predominant brain

285. *Mega Brain, New Tools and Techniques for Brain Growth and Mind Expansion*, by Michael Hutchison, 1986.

waves can be driven or pushed into new frequency patterns by external stimulation. In other words, the external signal driver or impulse generator entrains the brain, overriding the normal frequencies, causing changes in the brain waves, which then cause changes in brain chemistry, which then cause changes in brain outputs in the form of thoughts, emotions or physical condition. Brain manipulation can be neutral, beneficial or detrimental to the individual being impacted.

In combination with specific wave forms, the various frequencies trigger precise chemical responses in the brain. The release of these neurochemicals cause specific reactions in the brain which result in feelings of fear, lust, depression, love, etc. All of these, and the full range of emotional/intellectual responses, are caused by very specific combinations of these brain chemicals which are released by frequency-specific electrical impulses. "Precise mixtures of these brain juices can produce extraordinarily specific mental states, such as fear of the dark, or intense concentration."[286] The work in this area is advancing at a very rapid rate with new discoveries being made regularly. Unlocking the knowledge of these specific frequencies will yield significant breakthroughs in understanding human health.

Biomedical Instruments, Inc., an organization which markets electro-medical devices used to control or manipulate brain activity, was referenced extensively in Hutchison's book.[287] The power level needed to achieve a measure of control over brain activity is very small – from 5 to 200 microamperes – which is thousands of times less than the power needed to run a 60 watt light bulb. We are talking about very, very low power requirements. The trick for influencing brain activity is in the combination of frequency, power level

286. *Mega Brain, New Tools and Techniques for Brain Growth and Mind Expansion*, by Michael Hutchison, 1986, pp 114.
287. *Mega Brain Power*, by Michael Hutchison, 1994.

and wave form. What has taken place over the last two decades, and most particularly in the last several years, represents huge advances.

Research internationally found that the brain can be easily entrained or can be influenced to change states by external electromagnetic fields. These discoveries have provided new tools for both scientists and others. The new tools include electrical cranial stimulation devices, sound systems, light pulse systems and a large variety of other brain entrainment and feedback devices. Technological advances also have been applied to special monitoring and control devices which allow people to learn how to control and manipulate their own brain activity for beneficial results. Reports include relaxation, pain control, speed learning and memory improvements, among others.[288, 289, 290]

The work of Hutchison has yielded the finest descriptions of mind technologies yet put together. His book, *Mega Brain Power*, updates readers in an area which is changing so rapidly that the science is being formed faster than the applications can be fully recognized. Until some years ago, he published the latest research in a newsletter format. Back issues[291] discussed technologies for healing nervous system disorders, correcting attention deficit and hyperactive disorders in children, and curing drug and alcohol dependencies, among other things. Electromedicine of this type is emerging as one of the most exciting areas of medical research.

The research in recent years has extended to medical and psychological applications with startling positive results. Some

288. *Super-Learning*, by Sheila Ostrander and Lynn Schroeder, 1982.
289. *Mega Brain Power*, by Michael Hutchison, 1994.
290. *Mega Brain, New Tools and Techniques for Brain Growth and Mind Expansion*, by Michael Hutchison, 1986.
291. *Megabrain Report*, The Journal of Optimal Performance, Volume 2, Number 4, 1994.

of these results have been recognized by the United States Air Force and others for use in developing new weapons systems.

Dr. Robert O. Becker experimented in the early 1960's with ELF (Extremely Low Frequencies) by putting the signal on top of a DC current to carry the ELF signal. In other words, ELF rode like the passenger on a bus, maintaining its own integrity but being carried to its specific destination. Dr. Becker tested this concept by using an ELF, 1-10 Hertz (cycles or pulses per second) signal on humans resulting in an increased loss of consciousness among test subjects. The weak ELF alone had no effect, and the DC current had a significantly reduced effect without the combination. Above 10 Hertz with DC current, the effect increasingly declined until it was no better than with DC current alone.[292] What this demonstrated was that ELF, those frequencies which most effect human brain functions, could be manipulated externally with profound results.

Brain Biofeedback

Real-time brain biofeedback is another area of brain research which is beginning to provide great opportunities to many individuals. This area has yielded new approaches for gaining control of the mind. Through interactive computer technologies, it is now possible to monitor brain waves on a real-time basis, so that the individual using these tools can see what brain waves graphically look like on a computer screen, while the person is thinking.

There are several systems now on the market, with one of the finest being produced in the United States by American Biotech Corporation. Using these systems, a person can learn to manipulate his or her own brain waves in profound ways.

292. *Cross Currents, The Perils of Electropollution, The Promise of Electromedicine*, by Robert O. Becker, M.D., pp 227.

Some of the things which have been achieved include reaching meditative states (states of consciousness not otherwise reached without years of practice), increasing the recovery rates of brain dysfunction resulting from strokes or accidents, and improvements in children suffering from attention deficit and other disorders.

These new brain technologies are being used to help people with various problems. A number of researchers have shown the effectiveness of brain biofeedback in breaking drug and alcohol dependencies. The tools and techniques have been used for curing a number of mental disorders by teaching people how to create brain patterns within themselves, at will, to change their positive state of consciousness. Researchers found that within 30 to 60 days most people can learn how to create a number of specific mental states at will, and without the continued use of these machines. The machines serve as a bridge to new levels of mental control. The techniques using the computer interfaces can be explained simply as a system for teaching how to control one's own thinking in one's own way to gain specific results. It can be viewed as a small child learning to control his or her legs crawling, then walking and then running. We, as adults, are still at the crawling stage when it comes to mental control. These technologies are bringing to us opportunities to take greater control of ourselves through better control of our minds. The value of this technology cannot be overstated.

Altering the way we think, if we are individually in control of these alterations, can be a healthy thing. It is disturbing to realize that governments are interested in these technologies, not for beneficial individual uses but in order to gain increased control over populations they view as dangerous. These technologies offer both great promise and a high potential for abuse. Since the technologies are here now, all of the work in these areas should be open to public review.

This would help assure the preservation of the most fundamental right of humankind, the right to think in our own way.

The work being done in these fields has profound implications for all of us. The knowledge held by government agencies, under the guise of national security, would speed the understanding of these technologies and their positive implications would be realized. If the government's research were made public, we would have a significant advance into one of the most important areas of human development. The increased efficient use of the human brain would make it more likely that we could solve the very complex problems existing in the world, in addition to improving the quality of life for many people.

Chapter Thirteen

Tools for the 21st Century

Brain Entrainment
How does that work?

We'll start with the meaning of the mechanisms behind brain entrainment or Frequency Following Response (FFR). Affecting human behavior via the brain can be done in a number of different ways. We first must look at the predominant brain wave activities that are created within the brain and what we generally know from the sciences about them.

For review let's break it down into four areas, alpha waves, beta waves, theta waves, and delta waves.

Delta: We'll start with the deepest states of human consciousness. When a person is in a very deep sleep, they are in the Delta state, which is approximately 1-4 hertz (vibrations per second or pulses per second). If you looked at an EEG of brain activity you would see a lot of pulsing energy within this range of 1-4 hertz.

Theta: The next level up is called Theta. Theta runs between approximately 4-7 hertz (pulses per second or vibrations per second). This is when a person is in between awake and asleep, where they are consciously dreaming but still not fully awake. This is where the majority of 3-6 year old children spend most of their time. A look at the EEG of their brains would show that this is where the predominant activity is, which gives a clear explanation of why sometimes it's difficult for young children to separate the imaginary world from the real world. It is this same place where their brains absorb tremendous amounts of information. Think of three to six year-olds in terms of language skills and social behavior as an example.

Sometimes we prematurely push children into an early frame of learning when their brains aren't fully developed for academic studies. In Europe they usually don't even start children in school until they're seven years old, versus four and five year olds starting in school and preschool programs in the United States.

Alpha: Next is the Alpha stage, which runs approximately 7-12 hertz. This is where we are when we are in "the zone," when we are focused for learning and creative work. An example might be when an artist or an athlete is at her/his optimum performance. For accelerated or certain particular types of learning this is an extremely useful brain range to be in. It's where one would want to be for intellectual and creative work. This state can be reached at will with little training using several technologies.

Beta: The next level is Beta. This is where humans are actively engaged, thinking, learning and, at higher beta frequencies, displaying less controlled and focused behavior. At the same time there's a little bit more emotional content in terms of the intensity. As we reach high Beta we get into agitated states where we are not in as much control as we could be.

So if you're looking at children that function at a too high (Beta), or too low (Theta) they are often cataloged as attention deficit disordered. Often, as a result, we categorize children as learning disabled. And yet really what may be happening, in many instances, when we diagnose too early, is that the brain just hasn't fully developed to where the Alpha rhythms and the Beta rhythms tend to dominate the waking state. This is where we want to be as developed, productive adults.

The interesting early studies by Delgado and others was the idea that external fields, when they were coherent, or rhythmic and followed very specific patterns, could override the activity of the normal brain. This could be accomplished with pulsating radio frequency signals, pulsating electromagnetic fields, with oscillating or pulsating light, and could also be done with sound.

The human ear can distinguish sound, sound waves and sound itself, but we don't perceive the very, very low frequency sounds. Animals like elephants can perceive the low frequency sounds. Other animals that have large receivers (ears) can pick up low frequency sounds. But humans cannot. Very high frequency sounds are understood by other animals like dolphins that can hear up to 250,000 Hertz, ten times the range of humans. Artificial hearing technologies will one day increase the range of human hearing to exceed that which exists at any level in nature.

So what happens is that the brain, not being able to hear below a certain range, cannot pick up the low rhythms that can cause brain entrainment unless you can create a sort of cancellation effect. Here's what's been developed...

Robert Monroe, back in the late 1950's and early 1960's, discovered beat frequencies for entraining the brain. He found for example, that if you sent sound into one ear at 15,000 cycles per second, or vibrations per second, within the range of human hearing, and one in at 15,007 in the other ear, they cancel within the brain and leave a beat frequency of seven hertz, within the Alpha range as the beat frequency that the brain follows in the example. The brain locks onto the beat frequency, and this is called Binaural Beat. The concept was patented by Monroe and used to develop a number of technologies. The real effect of Binaural Beat is pretty profound.

If you look at this image taken from the Monroe Institute, you can see a normal brain on one side. The energy is distributed across the brain with one side in this case dominating, where the energy is focused and concentrated.

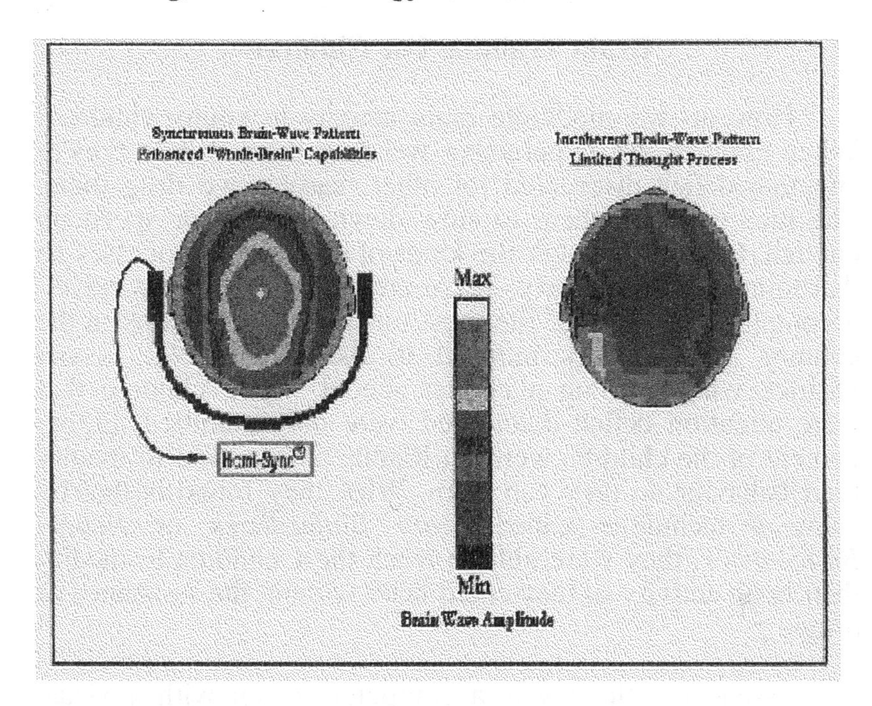

With Binaural Beats you create what's called *Whole Brain Entrainment* where both hemispheres of the brain, both the creative side and the analytical side, harmonize and work together. This is the ideal state of learning, which allows us to absorb information, taking it into consciousness and committing it to long-term memory. It also has very intense, suggestive effects on the human brain and subsequent behaviors.

More than 60,000 people were observed at the Monroe Institute utilizing this technology, and CDs were developed for creating very specific behavioral effects for enhancing human performance, not necessarily for military applications, but for civilian applications in a number of areas. The main interest of many was the idea of being able to enhance performance.

Brain Training For Children

How do we moderate brain activity in young children? When children are hyperactive we drug them. When they're attention deficit disordered for other reasons, we drug them. The idea is to get them to slow down enough, or speed up enough, to move through the kinds of intellectual processes so that things work properly. There are however, other ways to accomplish this. One of the most outstanding programs in the country was actually initiated in the Minneapolis School District, which set up a charter school where children that were attention deficit disordered were only admitted. Eighty percent of the children were on Ritalin, a drug used to modify their behavior so they can learn. What they found is that by using a technique called *Brain Biofeedback* or *Neuro-Biofeedback*, they were able to teach these children to modify their brain activity so they could learn without the assistance of drugs.

Essentially, there was a computer screen with a visual image to which a young child could relate, more like a game than anything else. Take a bouncing ball, for instance. The child would have a 16 or 8-channel EEG plugged onto his head, so he had all the points covered on the head. Now they make more modern looking systems that look like bicycle helmets without intrusive electrodes so kids will better tolerate them. It's a nice little arrangement where they look at the screen and as their brain hits the right range for that ideal state of learning (alpha 7-12 Hz), the ball will bounce on the computer screen. If the ball bounces in a specific range it indicates that the child's brain is in the ideal learning state. With this kind of feedback a great deal can be gained as the child learns to control their own brain activity in a period of several weeks. After this biofeedback training the children no longer need the computer hardware or drugs to allow them to focus their attention.

In thirty to forty one-hour sessions over ten to twelve weeks each child is able to actually be trained so that they can literally go into that ideal state of learning at will. In other words they don't need the biofeedback apparatus. It's like a training tool. It's the same as if you try and learn how to ride a bicycle out of a textbook. It's not going to work very well when you get on it the first time. But if you get on that bicycle and fall down two or three times, and learn how to ride it at five or six years old it is remembered because of the feedback of experience. If you never picked up a bicycle again until you were in your 20's you could immediately ride a bicycle again, having laid down the necessary learning tracks from the experience of actually riding with all of your sensory perceptions being engaged in the original learning activity. The tools developed for biofeedback allow a person to learn to control their own mental activity so that it fits the desired direction of our individual wills. Tools like these help to create fully engaged brains with all of their capacity. Brain biofeedback is essentially the same kind of focused learning allowing a person to retain those learning memory tracks required so that a person can reach these mental states on their own.

What did Minneapolis School District find out? After the first year, a large percentage of the student body was off of Ritalin and many other psychoactive drugs. They were able to move gradually back into regular education programs, not needing the special education resources that otherwise would be demanded throughout their twelve years of K-12 education.

Why is that important? Obviously for the individual it's apparent. But as taxpayers and people concerned about what goes on in our communities, special education is one of the biggest growing costs around the country. It's something in which other school districts should be actively engaged, and it's something that you as a reader could impact by becoming an activist team of one.

Anyone can help children and reduce costs in their community and actually begin to see technology applied in educational environments. Since Minneapolis started this program, a number of school districts around the country are beginning to duplicate that program and bring it into the mainstream. But it has taken twenty years from the discovery of the technology to actually getting it into an educational environment where it might be useful. If you ask people involved in children's issues, this is probably one of the most important matters facing children in public education around the country today. A reader of some of my earlier work in this area contacted the Minneapolis school district to get information on this program, and get it into her local school district. I received this message from her:

Dear Dr. Begich – I e-mailed you last October asking for more information on neurofeedback in the Minneapolis Schools. You suggested calling the MNPS

School District and asking about the charter schools program information on Neurobiofeedback or Brain Biofeedback. I was able to track down their program and thought you might like to have more specific contact information for future reference.

This program is part of a 501–C-3 called A Chance to Grow (ACTG) and their website is ***<www.actg.org>.*** *There is information about all ten of their programs, including their Neurotechnology program, on the website. One of their programs is a K-7 public charter school called New Visions School. At one point this school had a contract with the MNPS Public School District but my understanding is they are now independent of the public school district, and it took a while to find someone who knew about them. This school uses a number of innovative methods for helping kids learn to read, including neurotechnologies.*

Neurotechnology Services *is another program of ACTG. The current director of that program is* ***Becky Aishe <baishe@actg.org>. Her phone is 612-789-1236 x551.*** *The first director of that program, Michael Joyce, did pioneer neurofeedback in the public schools. Becky thought that was in '98 and '99, but this program first began in '91, according to the packet of information she sent me. Michael Joyce now works as a consultant to the program.*

They have done some additional pioneering work in the public schools using both neurofeedback and audio-visual entrainment (AVE), and have a training program for teachers or other professionals interested in these technologies. Becky said that almost 40 school districts are now using AVE under guidance from Neurotechnology Services. They found it simpler to teach use of AVE equipment than Neurofeedback equipment, so that is mostly what these schools are using. Becky made the comment that the larger the school district the harder it was to get new technologies in. Their 40 districts include districts all around Minnesota and in Mitchell, South Dakota. One of their school districts is the Menomin School District in northern Minnesota which I understand is a district with a predominantly Native American population. They are trying AVE on children with Fetal Alcohol Syndrome and I understand they are encouraged by the initial results.

*Another program of ACTG is the **Minnesota Learning Resource Center (MLRC**). This program is the one that provides training for educators, parents and professionals nationwide in the innovative programs of ACTG, including neurotechnology. They offer this training in August or September, and I hope that I may be able to take it. The Director of that program is **Nancy Farnham <nfarnham@mail.actg.org>. Phone 612-706-5519.***

The Development Director for ACTG is Gary Parker <gparker@actgorg>, 612-706-5515 or 612-840-0727 (cell).

It sounds like this organization has been at this for some time and is uniquely situated to be able to spread these technologies to other school districts or organizations interested in innovative educational programs. It sounds like there is some overlap in goals between ACTG and the Lay Institute, and some opportunities for collaborative work.

Thank you, Nick, for bringing my attention to this program in Minnesota. Even though neurotechnology is hardly embraced by the Minneapolis public school district as a whole, the use of neurotechnologies is flourishing, and ACTG and its various programs seem to be innovative, well-established and doing important pioneering work.

I hope this information is useful to your work, or to others interested in the subject.

Warm Regards,
Joan Bird

Biofeedback Continued

If you're looking for a practitioner of brain biofeedback in your area there is an organization in Denver called the *Biofeedback Association.* They're the only biofeedback certifying body in the country. Biofeedback can affect brain activity for children or adults, very useful in stroke recovery, addictions, behavioral issues, and attention deficit disorders.

Reducing stress is one of the most important health problems that has been intensified as a result of an increasingly complex culture based on technological change. Stress as a health issue is responsible for increases in both physical and mental health problems and death. Within the range of people's ability and on a more localized scale the *Antense*[293] (Alpha trainer) is a particular device I like, and it's just for relaxation and reaching alpha states. It incorporates what are called muscle tensiometers. In this case they are three small metal disks which pick up energy from the body and translate that energy in terms of muscle tension. When you tense up your muscles, there are actual changes in the electrical properties of the skin that can be measured. Sound is produced based on this signal in a pair of headsets. As a person listens to what starts as a high pitched tone and then reduces to a lower and lower pitch, the person begins to relax and then will reach Alpha brain states. The way the user feels, as amplified by the

293. Ibid.

sound tone feedback helps the user eventually learn to control his brain on his own without these kinds of tools. This is where biofeedback and mind training together can be very effective.

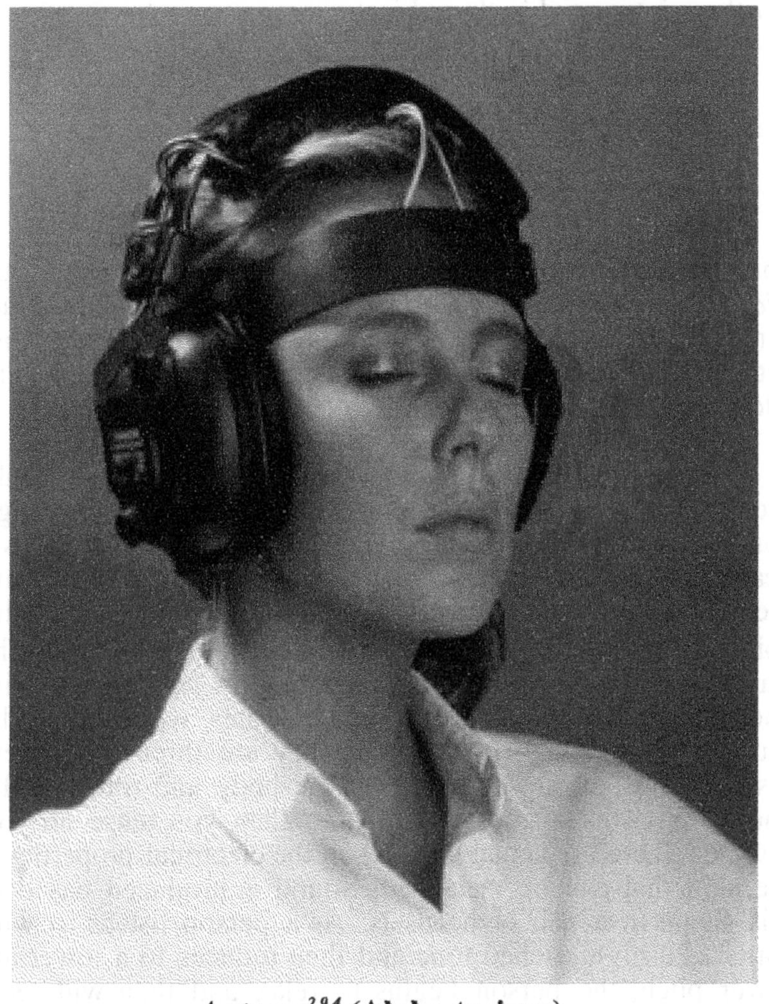

Antense[294] **(Alpha trainer)**

294. See at www.earthpulse.com

Kinesiology

Let's use the concept of kinesiology as an example. This is the idea that certain materials or energy interactions have an effect on our energy flows in the body with a direct impact on muscle strength. The system of measuring muscle strength discloses information to the kinesiologist about a person's health. If you don't know what the word is you may remember the image of somebody holding a substance in one hand and the kinesiologist pushing the other arm down to see how much muscle resistance there is. For example if you hold aluminum in one hand and it has a detrimental affect on you, your other arm will weaken. Or if you hold something in your hand that gives you strength, the other party can't push the opposite arm down when held out from the body. This method is "subjective" because you don't really know how hard the kinesiologist is pushing. Using muscle tensiometers you can wrap contacts around the arm and actually measure muscle tension so it is objective, not subjective. In other words by using a needle or digital readout you can see exactly what is going on with the muscle.

With the *Antense* device's muscle tensiometers there are three metal pads that fit into a headband. You've probably noticed when someone is stressed, they wrinkle up their brow. When they're calm, those muscles are very relaxed. Using this device, as your muscles tense you hear a very high-pitched sound in the headset, and as you relax the sound diminishes. That is your feedback, your interaction. Accomplishing this without the feedback is very difficult. Some people spend twenty years learning how to meditate. Why does it take so long? Because when they hit that ideal zone, it's only for that few seconds. And they remember, "Ah, that's what that feels like." Each time it is a little easier to get there. Eventually they get to that place very efficiently. Then they want to dive deeper. In those twenty years they learn how to dive deeper and recognize those signals.

It has been proven that people can reach the same mental state as the 20-year meditator using neurobiofeedback and brain biofeedback. Practicing once a day for an hour, you can learn how to drive to those slower brain states in 30 to 60 days. That is phenomenal and isn't done anecdotally. It is accomplished using many trials showing how you can drive the brain activity down, and then learn how to modulate your own brain activity at these levels.

That's key in a busy, hectic life; being able to learn how to immediately relax, reduce stress levels, and recognize relaxation in terms of the general sense of the body and being able to get there quickly and efficiently, even in the middle of the day with an intense schedule. It's nice to be able to create an environment where you can relax and learn how to work with your own physiology to accomplish what you want instead of using drugs.

Thought Stream

The *Thought Stream* is also a biofeedback device that actually measures skin resistance. A probe fits on the palm and if there is a little sweat, you can make a good contact. The device has a headset and when turned on it gives you an auditory signal so you can close your eyes. You don't have to look for a visual signal, you can relax. It's the same basic idea; the sound starts very high up and then drops to a lower pitch as you start to relax. When you turn it on you'll first see a light signal, which will set itself for energy state and skin resistance. The objective is to get the light to go from red to green. It will go from red to orange and start to work its way down as you relax. It gives a visual signal at the same time there's an auditory signal. If you wear the headset, it will tell you how relaxed you're getting by the pitch of the sound. You can adjust the sensitivity and volume on the instrument to get into more relaxed states. This device is a little more complicated than the *Antense* headset. Of the two, the headset (Antense-Alpha trainer) is the simpler technology to use.

There are more sophisticated brain biofeedback instruments, but they can run into thousands of dollars. The *Mind Mirror* is one, which actually has light bands that go across a visual screen for both hemispheres of the brain. So a person cannot only slow the body down but also make sure both hemispheres are basically energized at approximately the same energy state. A person can maintain that optimum brain balancing performance at the same time you're relaxing or going into those particular mental states desired.

Biofeedback teaches you to do it on your own. Light and sound instruments drive you there. For people with sleep disorders, light and sound devices are extremely effective for getting to sleep because they immediately entrain the brain into those relaxed states. The point is there are many devices that have specific uses depending on what is being sought.

Hemi-Sync & Robert Monroe

When you look back on Monroe's work, we can see he was doing some interesting things. He was trying to stimulate the brain, not just for learning applications but also layering the information over average brain wave patterns. Another thing that Monroe and others discovered is that you could combine Binaural Beat with flickering light. As an example most people think of the television set as a light radiator; it's pretty harmless. If you get back ten or fifteen feet as most of us do, and you watch the television set, it's pretty safe. It's what is called *non-ionizing radiation*, radiation that doesn't generate a lot of heat. You're far enough away from it. As the energy spreads out, the density, or concentration of energy decreases dramatically. This explains why on a television set when you get real close and put your hand on the screen, you can feel the energy radiating. But as you back from the screen it gets less and less dense fairly rapidly; within just a few inches you feel nothing at all.

If you think about radio frequency energy being broadcast from a radio station so you can hear your favorite song, when it starts out and you're close to the station, you hear the signal clearly. The further you get away, the less dense that signal is and the weaker the signal becomes until the music cannot be heard at all.

Regarding energy oscillators and light in particular, think back to the late 1990's. There was an incident in Japan where 600 children watching a television "Pokemon" cartoon actually went to the hospital with epileptic seizures.[295] What caused those epileptic seizures when light is supposed to be harmless, particularly that coming off of a television? It was determined it was caused by the flicker rate, which is the frequency in which a certain segment of the TV program flickers or pulses. It flickered in what's called a *window frequency*. A window frequency code is like a switch that when it is hit it causes something to happen. We could compare it to a radio station sending our music that gets received from our car radio. The car radio works when we dial up a radio station and hear the sounds. In between you get static, no signal, no music; but when we hit the right place on the dial we get clear resonance between transmitter and receiver, and get a nice clear signal. The same is true within our bodies and within our brains. Most energy is "static between the stations"; there is no resonance – nothing happens. But when you hit that "window frequency" you can

295.December 17, 1997 Pokemon, MSNBC Staff, Associated Press and Reuter contributed to the report.

trigger a cascading chemical reaction within the human body and brain. In that particular case in Japan it sent 600 children to the hospital after being exposed to energy fields otherwise thought to be harmless by government regulators. Many other brain states can be induced because of an increasing number of known responses to specific frequency coded signals that can be carried on any carrier of energy transfers.

Capitalizing on this knowledge, people have developed technology that focuses in on those window frequencies for enhanced human performance, whether it's to cause you to slow down, relax and go into those deeper states of sleep, or whether it's to perk you up, to gain energy so that you can move forward into the day with a lot more vigor and a lot more energy, utilizing this technology to deliver information, putting the brain into that ideal state for learning, and then bringing the information into memory in a way that the brain can absorb and actually be able to use it in a more effective way in the future. Other places people tend to want to experience are the deeper states of meditation, where we open our consciousness to our more creative capacities.

These are essentially the ways in which these technologies have evolved. And they have evolved in a number of different directions, such as light and sound devices and electro-cranial stimulation, using electromagnetic fields to stimulate the brain. There is also the concept of biofeedback, being able to get a signal in and being able to understand that signal in a way that causes us to slow down, or to be able to learn how to relax. Stress is probably the single biggest killer in the world today. Whether psychologically or physically based, stress is the root of many illnesses while also aggravating many other conditions. As our societies become more complex, the tools for relaxation become extremely important.

Electromagnetic Fields Affecting Bodies

Another interesting area is the idea of electromagnetic fields affecting our body. This is a source of much controversy and a lot of serious science, as has been pointed out throughout this book. I have personally reviewed many books and publications on bioelectro-magnetism, electromagnetic fields, powerline cover-ups, cell phones, etc. A lot of information has been published, some in the mainstream and some in the science community, and what is being reported today is fairly exciting. There are opportunities to enhance what we are as human beings, to become more complete and capable of reaching our higher potentials.

Consider the right brain – left brain arguments that went on in the 1980's when people were saying "women are more creative so they have one portion of the brain that dominates whereas men are more analytical so the other part dominates." If you really look at the EEG's and brain activity within human beings, you will see exactly that. But does that mean that one side or the other side should be more dominant? It's really a function of our education system. If you look at young children, you will see a more hemispheric balance between both sides of the brain, the creative and the analytical. As we go through formal education we tend to drift one way or the other and become channel-locked into what parts of the brains dominate. Ideally both sides work together, the analytical and the creative. The idea of enhancing performance on both hemispheres is the objective of the science of controlling the mind for our own advancement.

Chemical Interactions

The chemical model was the dominant representation of the last century for understanding the human body and energy interactions within it. This was the idea that everything that happens within our physical health and within our minds is related to chemical interactions. What was really a big controversy towards the end of the last century (but increasingly less of a controversy in the beginning of this one) is the idea that what happens to us energetically has profound effects on our health and our psyche.

If we think about the body, not so much from the chemical compounds – to body cells – to body components – to body, but we go a little bit deeper, think about the body first energetically and then in this order outward: energy, atoms, molecules, collections of molecules, chemicals, body parts, and then body. Starting at the very lowest level, energy can have a profound effect in terms of how our bodies work.

This knowledge has been often ignored because of artificial dividing lines among the sciences. Often people who had strong math skills and capabilities in mathematics went into physics and quantum physics. Those who had other interests but had good science minds tended to go more towards chemistry, physiology and the health fields. That breakpoint segregated the knowledge in a way that was not wise. We lost the understanding of what was happening behind those chemical reactions, and how we might be able to manipulate chemical reactions, not with other chemicals, but by using energy itself.

As mentioned earlier the military commissioned a study in the mid-1980's, through the University of Utah, that resulted in the *Radio Frequency Dosimetry Handbook* published in 1985. This handbook was actually put together to look at radio frequency energy and its effect on our health and on our mind. What they found is that every major organ of the body, including the brain, heart, liver, and lungs, could all be interfered with or overridden by external signals in the radio frequency range within a very narrow bandwidth. In other words, again, looking at the analogy of the radio and dialing through the stations, when you have resonance, a correspondence, a harmony between the transmitter and the receiver, that's where the energy exchange happens, that's where the action is. For a radio station it means a nice clean sound. In the case of the human body it means the efficient operation of an organ or the interference with that efficient operation.

When you think about the body's energy and the way we relate to energy, this is a very important concept, and it goes back to the very ancient cultures based on the archeological evidence. Earlier I mentioned that around us exists 200 million times more radio frequency energy than the Earth naturally produces because of what mankind has added. When you include all the energy that man has added into the environment, it's a tremendous difference. The idea of being immersed in a sea of energy and not really sensing it is where most of us are today in terms of sensing the energy. Here's a way to see the difference. When power grids shut down, most people immediately notice how quiet it is because the refrigerator's not humming, the fan motors aren't buzzing in your computer, etc. If you think of the general state of your body when the energy fails, it's almost like you've exhaled, as if your whole body relaxes. Think about it the next time the power goes out. Or if you want to experiment at home, go to your circuit breaker and stop the power coming into your house (remember to disconnect sensitive electronic devices when doing this). Notice the difference as your body relaxes because your body constantly has to create equilibrium. A certain level of stress is created within our bodies to compensate for those external fields that dominate so much of our lives today.

In the room that I'm currently in I am surrounded by a 60-hertz grid, 60 pulses per second. This is in the high Beta range. This is an agitating range for the human body. So we're always in this constant state in the United States and Canada of 60 cycles, where we're always just a little bit on edge. And we only notice the change when the power goes out.

Schumann's Resonance

There's another way to notice the connection between energy and your body. What do most people do when we walk into our house? We kick off our shoes. Why do we do that? People say, "Well I feel more relaxed." Why do you feel relaxed? Think about the insulators on the bottom of your feet. You're separated from the ground. The Earth has a natural oscillation, or pulse. It's called Schumann's Resonance. It was discovered in Munich, Germany in the early 1950's and represents the pulse rate of the planet at 7.83 hertz, which is right in the middle of the Alpha range. This is the ideal state for human learning and creativity, and it happens to be the literal pulse of this planet. As we separate ourselves from that and lock onto those stronger energy fields, now that are 60 cycles and higher, it explains a lot of the general stress that we see manifesting itself in the industrialized world that doesn't exist in the less industrialized parts of the planet or in almost all of mankind's developmental history. That is an important consideration.

Discovering Acupuncture

A segment was run on the Discovery Channel a number of years ago about the "Iceman" that was found in northern Italy or southern Germany, depending on whose side of the argument you're on. Apparently he was right on the border. Melting out of a glacier was this person who died 5,000 years before. They discovered that he was carrying medicinal herbs and other things that we know about today, but we didn't know that they had this knowledge 5,000 years ago. As they were doing the autopsy on this man, they found a number of tattoos covering his body. It just so happens that one of those people in the autopsy room noticed that the tattoos lined up with acupuncture points and acupuncture meridians. So they decided to take a look and see if this person actually had illnesses or disorders that were associated with those acupuncture points and meridians. Surprising to those that were in the room was that they did line up with the diagnoses, proving the knowledge was there. These points absolutely correlated. So how did the knowledge get there, in Europe? Did it originate in Europe, or in China as most believe?

I'll relate another story and will tie the two together. A good friend of mine was a physicist specializing in the effects of electromagnetic energy on the human body, Dr. Reijo Mäkelä. During his early work in the 1970's he began to take a look at mapping the human body. What he discovered is that all over the human body there were places where our skin resistance differs. In other words where there were concentrations of energy versus no concentration when measured with sensitive tools. Reijo began to map these places. This was done at Queensland University in Australia. He found that as he brought people in and tested them, everybody manifested the exact same concentration points of energy. He began to map them in the front and the back of the body over a number of years.

One day a student came into his study and said, "Dr. Mäkelä, why do you have an acupuncture chart on the wall?" Dr. Mäkelä laughed and said, "No, no, this is my work. I've been doing this for two years, measuring skin resistance on the surface of the body, the electrical properties on the surface of the skin." The next day that student brought in a copy of an old chart. It was an interesting illustration. It was a diagram from the ancient Chinese. It went back thousands of years, but not 5,000 years, not the age of the European "iceman".

The questions were, how did the Chinese get the information into Europe? Or, how did the Europeans get the information into China? I will share my theory, and this is strictly a theory. I want to differentiate that from factual information. Theories are ideas that may help explain what we otherwise don't understand.

It is my *theory* that knowledge was held by sensitive humans called in some traditions "healers, shamen, and mystics" who could sense these fields of energy. Through the ages the idea that sensitive humans had hidden capacities to sense these fields has always accompanied all sacred texts and mystic writings. When we strip away all of the "background noise" of manmade energetic systems of every kind imagined, and new ones being created even as these words are written, perhaps we could all begin the sense these energy fields that science can now also measure. Rediscovering the quiet energy that drives the body, emotions and mind, that which we each are as individuals, is where science and the religion merge. Many religions, false prophets, mystics, shamen and others have demonstrated anomalous human potentials for healing, extrasensory perception (ESP), creating "miracles", telepathy, and other unusual phenomenon. Science seeks to understand how these things work and as we do we also gain our own potentials to stimulate these phenomena and enhance our own human potentials as created beings. The possibilities that are being discovered will continue to challenge the foundations of our belief systems as they adapt to the realities that our technologies are creating.

Pointer Plus

The use of electro-acupuncture and the science of energy medicine is beyond needles and increasing in power. There are many systems for reorganizing the energy of the body. The use of these devices is more acceptable to Westerners because of our aversion to needles. In addition the devices tend to be more accurate at hitting the points and more potent in creating sympathetic effects in the body as energy flows are stimulated.

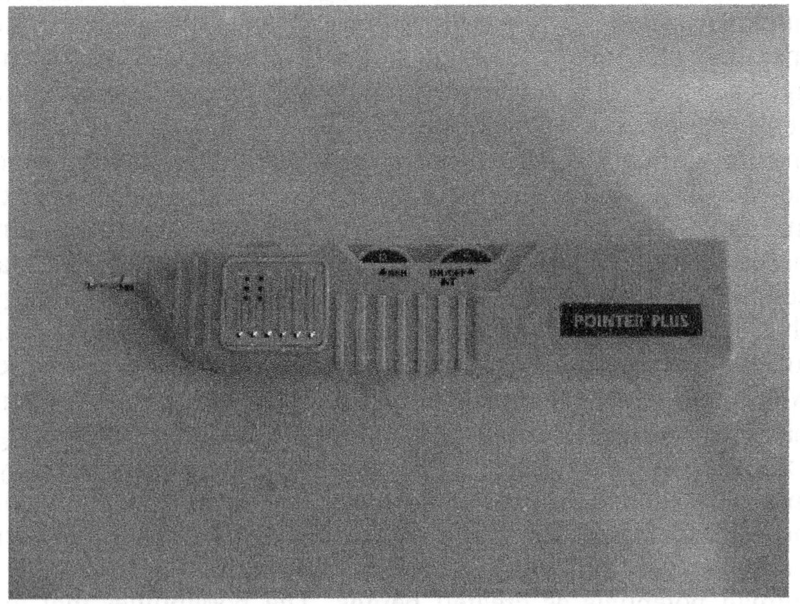

The *Pointer Plus* Electro-acupuncture Device

The *Pointer Plus* is an electro-acupuncture device that I have used for several years. It has two basic functions. Its first function is to locate the acupuncture points on the body accurately by measuring the differences in the electrical properties on the surface of the skin. In the other mode of operation of this device it sends energy into the body to stimulate the points it has located. When it is turned on, its power circulates through and a point can be easily located. When a point is found the device makes beeping sound and a flashing light. The more rapid the flash and even the sound, the more balanced is the detected point. By depressing a trigger mechanism, energy is actually sent into that point. It is perceived very lightly, almost like the dusting of a feather on the surface of the skin. It is a slight pulsing sensation. If it's more than that, the power is too high and it becomes uncomfortable. Once treated, the flashing light appears to be on, rather than flashing and the sound is steady.

This method of electro-acupuncture can be used instead of needling by sending just a little bit of electrical energy to stimulate a particular point. All of the points on the body can be found utilizing an instrument like this. This is sort of the modern equivalent of the 'sensitives' who could actually locate those points by their hypersensitive touch, or their electronic sensitivity, and be able to detect very subtle differences. These acupuncture points don't line up with the nerve bundles. They don't line up with muscle tissue. They don't line up with the circulatory system of the body in terms of moving the blood around. This is the fourth system, the energetic system of the body which is just beginning to be understood by medical professionals in the west.

Looking Forward

Energy-based medicine will be the medical science of the future. We need to look forward to this shift because in it we get a curative system of addressing health at a fundamental level. Unlike chemicals, which work very quickly, this type of approach to health tends to take a little more time, and it's not just the effect of affecting the energetic system of the body but also making sure the body has the basic nutrients through the foods we eat or the supplements we take, getting back to the things that some of us have lost in our modern diets.

We need the right building blocks in our bodies. Then we need to be in the right energy state to take advantage of those building blocks to build a healthy body and a healthy mind. That's what the future offers.

We're going to discuss some additional tools to get more of a practical look at how they work, what they do, and at some of the ways that they can be beneficial in our lives. So much of my work in the past has been centered on military activities, and the sinister uses of good technology. Technology is a two-edged sword in real terms. A truck can be used to deliver your mail, or it can be used by a bomber to blow up your house. So technology is not bad in and of itself. It's really the intent of the operator that matters. There are tools that deliver to individuals the power to take control over certain parts of our lives and enhance our own innate, natural abilities.

I mentioned light and sound in terms of the negative effects, about the Japanese kids that were watching a cartoon and had epileptic seizures. An important thing to note about light and sound devices, if you have any history of epilepsy, you cannot use these technologies because the light itself will trigger those kinds of seizures in people that are already very sensitive in that area. That's why physicians tells their patients not to go into bars or dance halls where strobe lights are flashing if they have a history of epilepsy. These individuals can, however, use the sound component, Binaural Beat. There have never been any reports of adverse reactions because it's strictly sound creating an entrainment effect for very specific types of enhancement.

Light and Sound

Light and Sound machines are essentially a signal generator, headset and a pair of glasses.

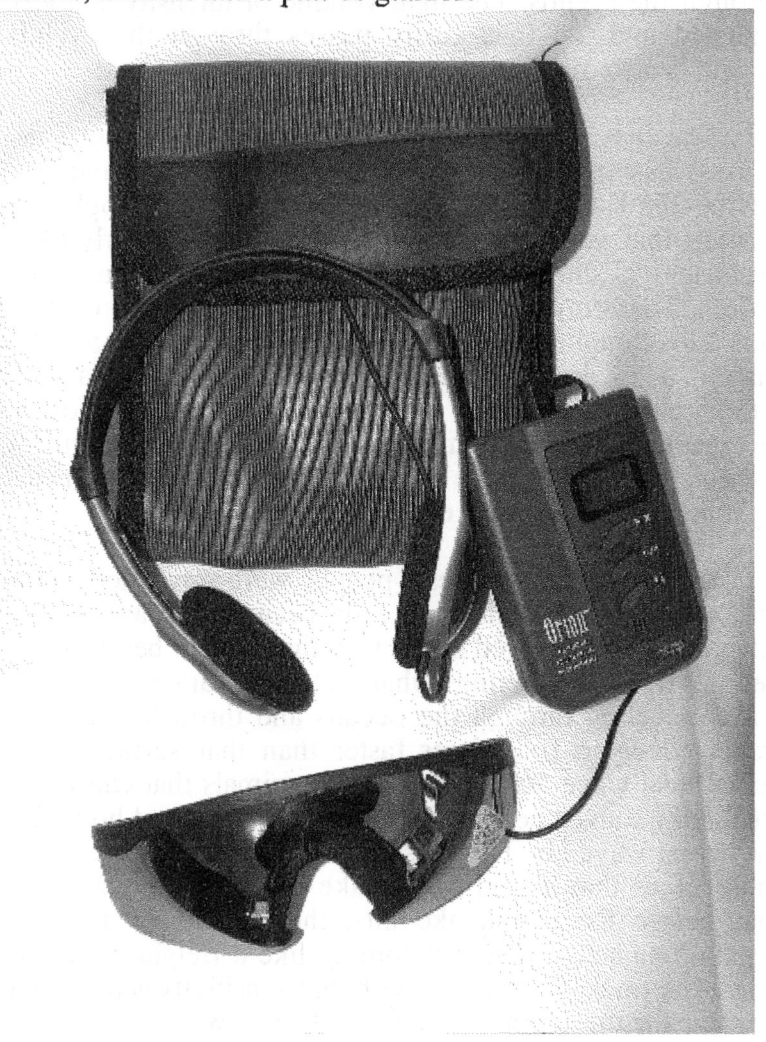

Orion Light and Sound Device - Now called Sirius

With light and sound devices you don't keep your eyes open. When the glasses are on the eyes are actually closed and, when the device is activated, the flickering light is seen through the eyelids. The light volume or intensity can also be adjusted and the light easily passes through the eyelids. A person is also more relaxed when eyes are closed.

The first step is to set a program. This particular device (Sirius) has more than twenty preset programs, and it's the device that I use predominantly. There are more sophisticated devices that you can self-program, but I particularly like this because it's simple. It has three buttons and an “on/off” switch. That's my three-button rule. If it's got more than that it gets too complicated for most people to use as a practical matter. This device is used primarily for four main areas, although there are twenty different programs. Those four areas are meditation, relaxation, accelerated learning, and enhancing creative work or creative capacity. Binaural beat and flickering light is used to create the effect.

As a point of interest, people were mystified after the December, 2004, tsunami in Asia because the animals didn't appear to have suffered in the way that human beings had, and people were wondering what actually happened. A sound wave traveling through the oceans and through land actually travels three to four times faster than that surface wave of actual tidal wave action coming. So animals that can hear low frequency sounds heard it coming at them several hours before the actual wave hit. They essentially heard the earthquake. For those of us that live in earthquake regions, most of us know that before the earthquake hits, there's usually an auditory signal. You actually hear it coming like a freight train coming down the track. So animals with that sensitivity actually pick it up. But for the brain of people to hear low frequency sounds we need this binaural beat, this cancellation effect.

In the case of light, the pulse rate matches the canceled sound beat so they are harmonized. This combination is quite powerful in creating the FFR effect. Essentially that's how light and sound devices work, with the flicker and binaural beat effects. As you watch the light flicker the brain entrains to the flicker rate, changing brain chemistry, altering consciousness and changing your mental state and your ability to either absorb information, relax, meditate, or engage in creative work. These devices are simple to use.

In learning applications this device (Sirius or Proteus) can be plugged into a sound recorder, CD player, computer or other sound storage device and allow a person to play back information while the device is activated. This way a person can create his own input of suggestions, or other information, he wants to enter his conscious and subconscious. This allows an individual to control his input in the same way we select other material to read or study.

The other thing that happens with light and sound devices is that blood chemistry changes. You can measure blood chemistry. You can actually see those changes within the blood workups, and you can essentially look at how your body reacts to light and sound. This is something that has been done. We published it in the *Earthpulse Flashpoints Volume 1 Number 6*, a sixty page booklet that breaks down four or five studies of light and sound devices specifically.

Electro-Cranial Stimulation

There are other devices for electro-cranial stimulation (ECS) to create brain entrainment. I don't care much for this method personally but will explain it a little bit although there are many ways to get energy into the body to have these very same effects. The reason I don't like this device is that it requires four contact points which are made with a headband that require contact gels. It requires moistened material when it makes contact with the skin so you get good electrical conductivity. What happens is essentially very similar to light and sound devices. There is an off and on switch and a switch for power levels. One switch actually sets the frequency, which ranges from a fraction of a hertz up to about 12-1/2 hertz. This covers all of those main areas that we're interested in, the Delta, Theta, Alpha and low Beta ranges. The device can be used for learning, concentration, and relaxation, essentially the same uses as light and sound, but in this case we're using electric current rather than light or sound to carry the required pulse rate or frequency code.

After placement of the four contacts a person feels across the cranium a pulsing, stimulating field, the electromagnetic field. The brain locks onto and begins to mirror the effect exactly the same way as the light and sound device. This device is a little tougher to use, but an interesting device nonetheless. I've not seen very many of these left on the market, mainly because of the inconvenience of use. But these are several ways to get the brain to oscillate in a different way at a different beat frequency.

Back To Electro-Acupuncture

In terms of the energetic system of the body, let's go back for a moment to the idea of electro-acupuncture. There have been other observations about energy and the human body. Back in the 1930's in Eastern Europe was the development of Kirlian photography. This is the idea of being able to photograph the energy discharging from the human body. My fingertips were photographed by putting my hands within an electromagnetic field generated by a Tesla coil and touching a photographic plate in the dark. The photographic film actually catches photon energy and light discharges coming off of the surface of the skin, in this case my fingertips.

There are people who purport to have Kirlian cameras often in fairs and new age shows. But they're not necessarily the true Kirlian photography. What they're doing is measuring skin resistance on the surface of the hands when you place them on a plate and those skin resistance differentials give you some variability so you get some energy patterns. With the original Kirlian cameras you would have to touch a conductive material, which actually loops through, so when you touch that photographic plate you actually get a bit of a shock. Your tendency is to pull away. It won't hurt you, but it's a little bit uncomfortable. So while you're getting that energy discharge you're placing your hands in the dark through a sleeve to where you're actually touching what is a photographic plate on the inside. The Kirlian cameras are interesting for demonstrating that there actually is this energy coming from the body.

Back to Hemi-sync®

Let's go back to the idea of hemispheric balance and brain entrainment. Again to remind the readers, the brain in normal state usually has one area or another that tends to dominate. Usually it's on one side of the brain or the other and the energy distribution across the brain is not so uniform. Yet when we use brain enhancement technologies, such as *Hemi-sync®*, we actually create a binaural beat. Both hemispheres of the brain begin to resonate and harmonize so the energy distribution becomes even. These CD's are each created for very specific uses with approximately 200 versions for different effects such as attention deficit disorders, stress, enhanced learning, meditation, speed learning, hypertension, pain management, creating OBE (out-of-body-experiences and other vivid imagery) and many other uses.[296] There are two separate kinds of sound tracks involved. One strictly uses tones and sound to create the effects and the other overlay words on

296. See www.earthpulse.com for demonstration of effect, research and full listing of all effects.

classical music with those tones and sounds. When you relax, the brain automatically tunes in, captures that binaural beat frequency and begins to respond as the CDs were designed.

For example, while using just tones and sound signals for learning, there are a couple of CDs where you can actually plug the sound signal in, put on your headset with *Hemi-sync*, and then read your normal material. At that point both hemispheres are working together. You're absorbing the information much more efficiently for language or any type of learning applications. It is extremely useful.

Virtually any auditory information that you can get can be used in conjunction with creating that signal. For instance you can plug in your college class lecture or a particular learning program that you're interested in. The low-end light and sound devices described elsewhere will allow you to jack in an external audio input, in other words that lecture or auditory information that you want to commit to memory or to learn.

Another way *Hemi-sync* is effective is when subliminals are used. If you have a multitrack stereo you can hear the words being said on *Hemi-sync* because they are not below the threshold of hearing. This is extremely important. You don't concentrate on the words, you concentrate on the music, which gets you into that relaxed state, and the words become the programming. The programming is important because words are symbols for thoughts and they mean different things to different people. Let's take the words describing a dog for example. Some people get warm fuzzy feelings when they hear the word "dog." Other people get gripped with fear because their experiences aren't the same. Word symbols have a lot to do with how we absorb the information. It is very important that those word symbols line up with our own belief systems. With any subliminal you should know exactly what is said and exactly how that information is delivered before you subject yourself to it. All these technologies put you into a super-suggestive state where the information flows past that normal filter that sifts out right and wrong, good and evil, the things that line up with our beliefs and the things that don't, and it dumps information directly into long-term memory. You don't want to inject information that conflicts with your normal beliefs. Always look carefully at your subliminals no matter who's producing them. Make sure they line up with your belief systems.

The Earthpulse Soundwave™

The next technology started with sound transfer devices, based on the frequency codes, that go back to the 1960's. The current devices use a sound signal and then conditions it for transit through the body. The current state of the art is the Holophon™ and the Earthpulse Soundwave™.

Normally how does sound enter the body? This thing we call an "ear" is really an array. It actually captures a soundwave as it hits the body and condenses the sound wave. It goes from the outer ear into the inner ear where it then hits the eardrum and creates a vibration. On the other side of that eardrum is a little bit of fluid. The vibration moves through that fluid, begins to vibrate three small bones in the inner ear, and then moves eventually to the eighth cranial nerve where an electromagnetic signal actually transits the nerve, and ends up in the auditory areas of the brain where the frequency coded signals are decoded and we hear sound.

If somebody jingles a set of keys while my eyes are closed I'll still be able to tell with fairly good accuracy the direction from which the sound is coming. If I distort the arrays, which are my ears, and close my eyes, it becomes much more difficult to isolate the direction of the sound. This device transfers the information to the brain via the frequency codes. The sound conditioning in the circuit is what allows the sound to be understood by the brain although transferred through a different pathway to the brain. That is where the signal gets transferred into a pattern of transmission that the brain understands. This is where researchers have discovered ways to replicate what the inner ear does. In other words, take a sound signal, condition it in such a way so you can bring it in, bypassing the eighth cranial nerve normally thought to be responsible for all auditory input into the brain, and you can actually perceive sound bringing it in through other parts of the nervous system, or perhaps that energetic fourth system of the body. The jury is still out on how the information gets into the brain's auditory processor. Recently there was an article in our Anchorage Daily News about sound that's actually perceived from the electrojet created with Auroral discharges. Some people hear it and some people don't. They don't understand the mechanism by which that Auroral sound gets into the head. I'd suggest that it's a very similar mechanism to the one I'm about to explain.

Earthpulse Deutschland advanced the technology significantly. An incredible inventor and friend in Germany, Robert Thiedemann looked at the old concepts of the dated technology, which included approximately thirty components on an electronic board in a box to create a sound that you could perceive literally in the center of the head. Robert looked at the old ideas, advanced them substantially, and created a much finer circuit. More than one hundred and seventy components are involved in the design of the circuit, which allowed him to get much lower and higher frequency ranges delivered into the brain. Unlike a standard auditory signal, the ears aren't involved because the signal is conditioned. You can use a standard CD player with Bach playing classical music and route it through this device's circuitry. About midway through the circuit there is an electro-optical coupler. This coupler changes the electricity to a flow of photons (light energy) and converts it back to electrons flowing as electricity in the body. He included the coupler in the design to act as a circuit breaker. This is necessary because although we're attached just to a CD player, if you hook this up to a home stereo or a computer where you're actually hooked into the wall with AC power, a person doesn't want power to arc across the circuit and into the body. At the other end of this device there are piezoelectric transducers. They have a metal contact point. When we make contact with these we actually transmit current through the body. If you listen carefully you'll hear a little bit of energy discharge. These have ceramic material on the back of the metal plates. As electric current hits that ceramic material, it causes it to expand and contract, which creates a slight drumming effect on the surface of the metal plate. That drumming effect then creates a sound that you can hear with the normal ear. If I actually place this on my temple and block my outer ears, the sound is internalized. The sound is actually heard as if it were originating in the center of my cranium. When you're out of balance energetically, you'll find that the energy will stay on one side or the other. Within about a minute of use, the energy centers of both hemispheres begin to beat together. You get exactly the same effect only it's created as a condition of use, not the input signal. In other words, any input signal creates this kind of balance in both hemispheres to whatever beat frequency the sound might generate.

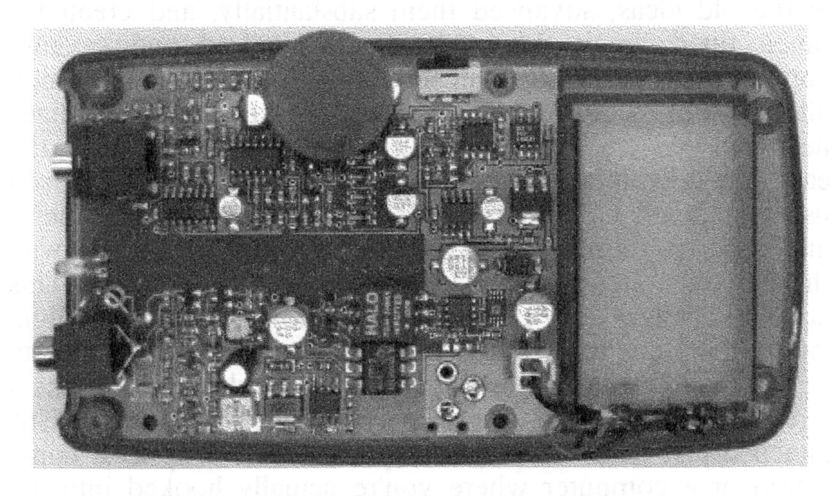

Earthpulse Soundwave™ Circuit

Where this has been particularly effective in Germany is the combination of Hemi-sync technologies together with this delivery system using the *Earthpulse Soundwave™*. The reason is a person is not just receiving sound through the normal auditory channel but also getting it through the new auditory pathways, increasing the inputs and stimulation. With this instrument you can split the sound output cable so the signal can also be hooked up a regular headset. You can then put on the piezoelectric transducers and end up with a three-dimensional sound. It's the same kind of difference that you would perceive by listening to AM versus FM where the sound, volume and quality fills out with FM, it fills out one more level by combining the *Earthpulse Soundwave*™ technology with a regular full range audio headset.

For brain enhancement and body balancing, where the *Earthpulse Soundwave* has been used. It also adds electrons to the body through pulse-modulated piezoelectric transducers, creating an energy flow through the body. Electrons are being delivered to the body. They enhance your overall energetic state. The device also balances the hemispheres. They also balance the acupuncture meridians, and it will show up. A week or so later a person's system will still be in balance after about a half hour of use, assuming that they don't experience some huge stress levels or traumatic event that throws off their equilibrium. Generally speaking, the *Earthpulse Soundwave* stimulates a whole energetic system balance, allowing for increased learning and enhanced performance.

This technology was developed for Earthpulse Press and Earthpulse Germany. We hold the U.S. patents. Our colleagues in Germany hold the international patents for the technology. We are just introducing this as one of the leading-edge learning technologies now available. Initially, as we field-tested the device we found that certain individuals with hearing impairments or hearing loss, particularly in the high frequency range, were able to regain that high frequency loss from this sound enhancement device. We also found that people using this could perceive sound signals above the normal range of hearing up to the carrier frequency of 40,000 hertz. If listening to a standard headset compared to high fidelity sound, it only goes so far. When listening with the transducers it jumps another level. You can actually catch the higher fidelity sound that otherwise you would totally miss on the audio input, whether it's music or a lecture. The combination of a regular headset provides the best range of sound. The signal is not strong enough for use as a hearing aid, however, we are considering the advances that are possible in talking this technology further in the future.

Chapter Fourteen

The Future

This book went through an extraordinary number of edits in order to clarify the writing as much as possible for readers. It was our hope that this book would be a catalyst for the debate that is sure to follow on the subject of *Controlling the Human Mind*. I consider this one of the most important issues in the beginning of the 21st century. The possibilities of being freed to our higher potentials or enslaved by those afraid of that potential will be decided by this generation.

The subject matter of this book went well beyond the beginnings of the credible research into hypnosis of the 1920's. Ideas advancing in the military think tanks of the world, for over 85 years, are now integrating as the "controlled effects" of modern militaries. Controlling the human mind and deceiving it into whatever thought, imagining, memory or other information delivered by any number of advancing technologies will become increasingly tempting by policy planners. These technologies will continue to become more sophisticated and advanced. One of the hopes in compiling the material used for this book was to stimulate the kind of ethical discussion that would lead to good public policy domestically and internationally.

The temptation to use the power of modern technology to further what some individuals might believe is "right and correct" must be avoided. The most basic of human rights has to rest on freedom to think without any uninvited invasion if there is to be any true individual liberty.

The recognition of this fundamental human right must follow a clear understanding of the existence of these technologies given the volume of material available in the open literature for those with the time to uncover it, understand it, and explain it in clear language. Moreover, the disclosure of the underlying principles that could improve the human condition must be reviewed, and where possible, applied to real health solutions.

As this book was being completed the next one was being planned and drafted. I have begun to work on a project with another author, Richard Alan Miller, on the subject of enhanced human performance. That book will be out in the early part of 2007 and will deal with the tools that are available to help accelerate the development of body, mind and spirit. We believe that the correct use of technology begins with the freedom to use technology as we individually choose as long as we do not impose our will on others without their consent. These technologies, because of their very nature, open the door to larger questions that will hopefully lead to positive change and advancement.

The idea that we are on the planet to self-actualize our highest and best potentials has always been imbedded into my belief systems. We are called to help others reach their potentials and, in so doing, we enhance our own possibilities as almost a "side-effect." The tools of the 21st century are not just about improving the mind but are about our underlying essence and the systems that regulate our physical form, our mental/emotional processes and our connection to everything else in creation.

We can enhance our human experience and I hope that as I conclude these pages I leave readers with a sense of hope rather than fear. It is my sincere desire that we can turn the direction of these technologies toward human service rather than as the tools for a "new world order" from which no person on the planet shall be free.

The next book project will expand Part II of *Controlling the Human Mind* and we hope it forms the framework for increasing access to the advances of technologies that will meet the modern needs of people. The tools that reduce stress, enable any conscious state to be learned quickly, and increase one's mastery over the body and mind will be discussed in-depth. We will draw from many seemingly disconnected bodies of knowledge to educate, illustrate, demonstrate and make available some of the technologies that might help shape the 21st Century. This will be the century of the mind, and as such, it will be a century where we open the gateway to a deeper understanding and appreciation of the Spirit of mankind.

What Needs to be Done

There is a great deal that can be done to address the issues raised in this book. The process of public education is the focus of our current efforts. We are also intending to initiate several projects that will continue to build interest in these matters in the hope that the wisdom that flows from open and honest dialogue will alter the directions we would otherwise be headed. Readers can assist us in several ways:

1. When you have completed reading this book send your copy to your Congressman, Senator or Parliamentarian and request a full inquiry into the state of these technologies. Include a hand written note expressing your opinion of these technologies.

2. Consider becoming part of out activist network by signing up to the flash list at ***www.earthpulse.com***. We use this list as a way to contact people about projects, lectures, interesting updates and information about our research. We are also going to be using this site to coordinate political activity on technology issues. This is a private list and is not shared with any other organization.

3. Stay informed by reviewing the ***www.layinstitute.org*** website which contains the research material we draw upon. We are loading new materials to this site periodically. The site will be coordinating educational and other technology initiatives that will be useful in engaging public awareness of many new breakthroughs in science.

4. Over the years I have presented in hundreds of lecture forums, radio, print media and television interviews around the world. We encourage people to get our information to those who can help get the word out on these topics. Because of readers and radio listeners we have been able to greatly multiply our outreach. This is an area where everyone can help. Call your favorite talk show host, organization that sponsors lecturers or any other public media and have them contact us or use the information on our websites.

5. In many instances people familiar with our work, and the many subjects we cover, send us news articles and other reports that provide additional references that support our public education mission. When you see interesting material please forward it to us at the address in this book. We will use it in our future work and make it known to the public as resources and opportunities allow.

6. Consider supporting our activist mission by purchasing our books, DVDs and other materials as part of your personal gift giving occasions. Our resources go back into the work we are trying to accomplish and we appreciate all of your support. Over the years we have been able to accomplish much and hope to continue in the years to come.

7. Follow your passions in life and be an "activist of one" on the issues that are important to you, no matter what those issues might be. We are part of the most powerful network ever created – the ultimate *"organic internet"*, if you will. It's the Human race created in the image of the Creator. We are *designed* to create and be stewards over this earthly reality – so let us create with care, with respect and consideration for the essence of life, and be liberated to our individual potentials.

Appendix

The Lay Institute on Technology

The Lay Institute on Technology Incorporated was conceived and founded by Ms. Dorothy Lay of Dallas, Texas. In was established as a Texas nonprofit corporation and has filed with the United States Internal Revenue Service for recognition as a Private Foundation under section 501(c)(3) of the Internal Revenue Code.

Ms. Dorothy Lay is the daughter of Herman and Amelia Lay, also of Dallas, Texas. Ms. Lay's father was the founder of H. W. Lay Company in 1938. It was merged in 1961 with the Dallas-based Frito Company forming Frito-Lay, Inc. where he served as CEO and Chairman. In 1965, the company merged with Pepsi-cola and was renamed PepsiCo, and he was elected Chairman. Herman Lay served in many corporate capacities throughout his life until his death in 1982. The Herman Lay story is the bootstrapping story of a great American business leader who, from humble beginnings as a salesman, helped create one of the most dynamic companies in the world. Mr. and Mrs. Lay instilled the values of hard work, southern hospitality, personal character and integrity and community purpose in their children.

With the passing of Herman Lay a great deal of responsibility was transferred to his heirs and the values he instilled in his family continued to be expressed in their work in the communities in which they live. Dorothy Lay has been active in education, the arts and various community activities throughout her life. She began to take a particular interest in the direction technologies were heading and the impact of technology on people, our institutions and the planet itself. Her interest in technology and the need for increased public awareness of technologies and their potential impacts was the primary motivation for forming The Lay Institute on Technology, Inc.

In early 2004, Ms. Lay contacted Dr. Nick Begich to discuss the possibility of serving on the Board of Directors of the organization. During the summer and fall of 2004, the work of forming the Institute was completed and some initial activities outlined for the organization.

The purposes of the Corporation are to research, explore, study and educate the public, by publication and by public forums, in the following areas, to-wit:

I. The effect on life and society of technological advancements on the longevity of human life;

II. The creation of social and cultural institutions to meet the needs and define the directions of mankind where life spans and the presence of disease are effected by technology;

III. The effect and distribution of technology on the quality of life for all mankind;

IV. The moral and ethical issues of technologies which affect the quality of life for all mankind; and

V. New societal models and paradigms to meet the need of future generations from the effects of technology.

Bibliography of Technology

The initial reference material for the www.layinstitute.org website was compiled over 20 years by Dr. Begich and his associates. The articles and papers have been culled from thousands of documents reviewed over those years. It was realized that the indexing, and effort to collect these materials, formed a corpus of a research library that would be of great use to those interested in technology in the future. Making the information available to researchers and others interested in these topics is now possible.

The bibliography will be built on a continuing basis. The current content represents a small amount of the material we have collected. We have over 20,000 documents yet to be collated, abstracted and uploaded into this system. It is hoped that this material will serve to educate and enlighten this generation on the impacts and deep change technology presents. Moreover, with the transfer of these materials to the Lay Institute on Technology, Inc., they will be enlarged on a continuing basis by both adding newly uncovered materials and continuing the effort of indexing and organizing the unclassified materials remaining from the initial transfer of data.

As these records were indexed they were given individual identifying file numbers that began with the letters EPI followed by a number. As we researched and collected materials they were filed sequentially, assigned a number, and a bibliographical reference file was created to identify such things as title, author, date, place of publication and other reference information used to cite the material. We then created an abstract statement about the citation. The abstract was either a written quote from the article that would encapsulate its content, a written summation of the article; or, a lengthy direct quote in the event that the document was a public domain article or paper. We attached the EPI number to each footnote whenever we published books or articles which used the materials and this allowed us ready access to all master file documents.

Many of the documents that are contained in our master files are copyrighted so they could not be duplicated on this website. This is why we used limited quotes and new summations in creating the abstracts with the exception of material considered in the public domain.

How to Use the Bibliography and Abstracts

We have set this site up to be random access in the sense that the classification of material is not by subject but just entered into the database on a continuing basis. In this way the information can flow into the site with minimal classification error in that every word is searchable. If a researcher has a subject, author, publication or other search criteria it can be placed in the search engine and all records related to it will be sorted and identified. The user can then look at each abstract and bibliographical reference and locate the material with greater ease. As time goes on we will be adding much of the public domain material to this site as well as articles where permission to republish is granted.

We will also continue to evolve this site to increase its utility to all users. In this effort all users are urged to contact us with suggestions that might improve this site.

The Lay Institute on Technology, Inc.

How Can You Help?

1. Please forward articles, research materials, government documents or other materials which you feel might add to the work we are doing. This is very important and has helped a great deal over the years in providing information and leads useful in our work. Send such data to Dr. Nick Begich, Executive Director, P.O. Box 201393, Anchorage, Alaska 99520, for entry into this index.

2. Contact your favorite talk radio and TV hosts, magazines and other media outlets and let them know about our work and ask them to consider having us speak on the various issues we cover or provide data for their use informing the public on these important issues.

3. Contact speaker's forums and encourage them to have us lecture in your area. We have been able to personally appear in numerous forums when sponsorship has been arranged. We are reserving lecture time each month so that we can meet these requests.

4. Consider volunteering to help in our various efforts. If you are in the Anchorage area, we are in need of volunteers to work with us in our indexing project and other activities. Contact us at:

www.layinstitute.org

Resource Guide

A FREE CATALOG of many of our products and books is available on request.

1.☐Mind Control: A Brave New World or Enhancing Human Performance - DVD format video. Manipulation of the mind, emotions and physical health of people through new applied technologies continues to draw the attention of military planners around the world. The presentation provides in-depth information, demonstrations, background and forecasts of development of these areas of science as they affect our society and individual freedom. Enhancing human performance or controlling human outcomes will be the challenge of the century raising serious questions on the ethics of the science of mind control. Approximate running time 1.5 hrs. The DVD is $24 including US shipping or $28 internationally.

2. Technologies in the 21st Century - DVD format video. Technologies that will transform our lives are advancing rapidly raising serious questions about privacy, safety and their proper uses. Military planners and others are attempting to use these breakthroughs to create a more directed and controlled society while ignoring the positive applications of many of these new discoveries. This video delivers an overview and updates on the HAARP material, cell phones, privacy, underwater sonars and other areas of technology. The DVD includes discussion intended to stimulate debate and while educating the public on these pressing issues of the 21st century. Approximate running time1.5 hours. The DVD is $24 with US shipping or $28 internationally.

3. HAARP - The Update: Angels Still Don't Play This HAARP: Earth Rising Series Volume 3. Based on the best selling book, Angels Don't Play This HAARP, narrator Dr. Nick Begich presents a compelling discussion of one of the important military advances of the United States Government. The technology is designed to manipulate the environment in a number of ways that can jam all global communications, disrupt weather systems, interfere with migration patterns, disrupt human mental processes, negatively affect your health and disrupt the upper atmosphere. The U. S. military calls this new zapper the High-frequency Active Auroral Research Program or HAARP. The rest of the story is revealed in the patents, technical papers and other documents that continue to emerge regarding this project. Begich has presented on the subject as an expert witness for the European Parliament, Committee on Foreign Affairs, Security and Defense Policy Subcommittee on Security and Disarmament, GLOBE and others. Approximate running time1.5 hours. The DVD is $24 with US shipping or $28 internationally.

4. Earth Changes - The Ripple Effect: Alaska Sounds the Alarm: Earth Rising Series Volume 4. Arctic and Antarctic regions of the world are being profoundly affected by earth changes. The arguments over manmade or natural causes continues to be debated but what can not be denied is -Things are changing regardless of the cause! This DVD addresses many of the issues facing the all of us as residents and stewards of planet earth. The narrator presents his personal experience and knowledge combined with the latest research on these subjects. Climate changes, earthquake increases, pole reversals and pole shifts are discussed using Alaska as an example of the changes now taking place. In addition energy policy and alternatives are discussed as well as some of the economic issues related to earth changes. Energy policy issues are discussed in terms strategies that might be applied to national and international energy matters.Approximate running time1.5 hours. The DVD is $24 with US shipping or $28 internationally.

5. Angels Don't Play this HAARP: Advances in Tesla Technology is a book about non-lethal weapons, mind control, weather warfare and the government's plan to control the environment or maybe even destroy it in the name of national defense. The book is $19 shipping in the U.S. or $24 internationally.

6. Earth Rising - The Revolution: Toward a Thousand Years of Peace, by Dr. Nick Begich and James Roderick. This book is a book about the impacts of new technology on humanity. The book is footnoted with over 650 source references spanning fifty years of innovations. The book includes advances in the Revolution in Military Affairs, mind control and manipulation of human health, non-lethal weapons, privacy erosion and numerous other areas where technology is impacting mankind. The conclusion of the book contains an overview and series of actions that could be initiated which would return us to a more civil and open society. The book is $23 including shipping in the U.S. or $29 internationally.

7. Holes in Heaven - THE DVD. This documentary film narrated by Martin Sheen explains both sides of the debate surrounding the HAARP issue and includes the military, scientists and those opposed to the technology. This video is $29.00 Air Mail in the U.S. or $33.00 internationally.

8. Earth Rising II: The Betrayal of Science, Society and the Soul, by Dr. Nick Begich and James Roderick. This book is a book about underwater sonar, cell phones, energy weapons, implant technology, privacy and other emerging technologies. The book is footnoted with over 300 source references spanning fifty years of innovations. The conclusion of the book contains an overview and series of actions that could be initiated which would return us to a more civil and open society. The book is $23 with shipping in the U.S. or $29 internationally.

9. Secrets of the Soil, by Christopher Bird and Peter Tompkins, is the most important book ever published on new ideas for more productive agriculture. Faster growing rates and greater yields without using petrochemical based fertilizers is revealed in this incredible work. The book is of great use to home gardeners and commercial growers alike, an important book for all private collections. Available shipped in the U.S. at $24.00 or internationally for $30.00.

10. New Sound of the Earthpulse Soundwave™! Wait listing only. The most advanced sound transfer technology available anywhere in the world is currently being beta-tested in Europe for introduction to the United States. This new technology will allow for the transfer of sound directly through the body by-passing the normal hearing mechanism. For more information on this technology contact Earthpulse Press at the above address. This technology has been developed with superior technology created in Europe in 2001-2006. The technology far surpasses the performance of earlier sound devices in several applications by integrating modern circuit designs with 21st century engineering. The technology delivers sound energy through a biologically compatible signal. This technology will be available for accelerated learning applications, body system balancing and trials in sound enhancement applications.Estimated list $495.00 plus U.S. shipping $20.00 and $35.00 international. (wait listing only until fall 2006)

11. Light & Sound Technology - SIRIUS. Our new *Sirius*™ is the best value for money of any light and sound system on the market today. It includes a number of features normally found in machines costing much more. *Key features of the SIRIUS system:* ***a.)*** 21 programs for relaxation, learning, increased energy, visualization, sports performance, and more;

b.) Unique random session creates a different session every time it runs; **c.**) Manual control of pulse rate in tenths of a Hertz, from 0.1 to 40.0. Use this to create precise stimulation protocols for your personal exploration and special applications; **d.**) ColorPulse™ decode circuit converts rhythmic sounds (the "beat") from your cassette/CD/MP3 player into pulses of light. Turn any source into a synchronized light show!; and, **e.**) AudioPulse™ decoder and built-in microphone converts any ambient sound into a lightshow. ***Note:** System requires 3 x AA alkaline batteries, not included.* $133.00 USA/Canada or $145.00 International including shipping.

12. ***Electro Acupuncture - Pointer Plus*** is a hand-held, battery operated acupuncture point locator and stimulator. Pointer Plus replaces traditional acupuncture needles with a mild electrical pulse which can be easily self-administered. Use of Pointer Plus, like acupuncture, depends upon accurate identification and stimulation of the body's low resistance points which are located near the skin's surface and are a key mechanism for the natural relief of pain. Pointer Plus' sensor/electrode measures the difference in electrical resistance between a low resistance point and the surrounding skin. Precise location of the point is indicated by a buzzer and a flashing green light. Push the applicator button and a mild electrical pulse is applied through the same sensor/electrode for 15 - 20 seconds. This procedure is repeated for each of the points indicated for a specific need. Great for minor aches and pains and overall energy balancing. Unit is adjustable for voltage and point locator sensitivity. Uses a single 9 volt battery and comes with a sturdy plastic case. Electroacupuncture is in use around the world for numerous applications including: pain relief, energetic healing, general energy balancing, headache, backache, stress reduction, muscle aches, sciatica, stress and tension headaches, sports injuries, muscle pain, tendon pain, joint pain, tennis elbow, tinnitus, dizziness, asthma, hay fever, sinus problems, and emotional related problems. Purchase of Pointer Plus is non-refundable. It is assumed the purchaser has an adequate understanding of acupuncture techniques to make effective use of this instrument. $153.00 in the USA/Canada or $160.00 International including shipping.

13. Alpha Trainer - Antense® Biofeedback anti-tension device designed to combat stress in the privacy of your own home. The Antense antitension biofeedback device is a simple yet effective way to eliminate stress by helping you develop a greater sense of body awareness while in the comfort of your own home. Optimal for personal health and fitness. Using the scientifically proven principles of biofeedback, Antense makes you aware of how stress changes your body and how to better control these changes. Your body indicates its stress level by muscle contractions in the forehead, neck and scalp. Using space age patented technology, Antense measures these contractions and instantly converts the signal into a pleasant tone-pitch proportional to the level of muscle tension in your body. By listening in on the tension in your body you can control the pitch using the tone as a guide. By learning to feel how stress changes your body, you will quickly understand how to release that tension and stress and learn how to relax at will. Enables increased alpha and theta activity with use. This comfortable leatherette headset has an adjustable soft belt with dry EMG electrodes, a sensitivity control, adjustable volume and comes with a handsome storage case. (requires a standard 9 volt battery) *$174.00 USA/Canada* or $184.00 International including shipping.

14. Eighth Element™, is a custom blend of more powerful and potent hybridized Cordyceps sinensis. Radically increases ATP Production, Stamina and Endurance, and Oxygen Absorption - Proven to Increase Cellular Energy by up to 28%! For thousands of years, Cordyceps has been highly valued in China as a tonic food and herbal medicine. In nature it is rare, very expensive, and subject to contamination. Today, a potent strain of Cordyceps, CS-4, is grown in a controlled environment, under sterile conditions, making it safe, affordable, and available for use as nutritional therapy for many conditions. As with the other medicinal mushrooms, the anticancer effects of Cordyceps appear to come from its polysaccharides, as well as its sterols, lipids, nucleosides and especially in the de-oxy nucleosides, which have been found in no other source in nature. Many studies have shown this mushroom's ability to stimulate and modulate the immune system and increase red blood cells. As an immune modulator and adaptogen, Cordyceps was shown to boost depressed immune function, but not enhance a normally functioning immune system to the same degree. In published research from China, Cordyceps was reported to have anti-tumor activity against lung cancer in both mice and humans. In a number of human clinical trials and in Traditional Chinese Medicine, Cordyceps has been shown to improve liver, kidney, cardiovascular, and respiratory functioning. Since these organs and systems can be adversely affected by cancer and its treatments, Cordyceps can contribute to overall health, stamina, strength, and endurance besides having possible anti-tumor effects. ***Eighth Element™*** 60 each 500 milligram vegi-caps per bottle. *$24.00 USA/Canada with shipping included or $28.00 International including shipping.*

15. First Elements™. *Silica Hydride* powder propretary blend. We are introducing a new dietary supplement line called Earthpulse Nanotronic Nutrients for use in overall health maintenance. First Elements was formulated in Germany and outperforms similar silica hydride products. The nanotronic foods researched by Earthpulse and its affiliates in Europe has resulted in a number of new supplements being formulated and selected over the last six years. First Elements provides energy and, over a week's time, allows the body to return to a more alkaline healthful condition. Similar products sell for more that twice the price. Greater effectiveness and quantity for half the price - 60 each 500 milligram vegi-caps per bottle. *$24.00 USA/Canada shipping included or $28.00 International including shipping.*

Disclaimer: The supplements and devices described in these pages are not medical devices and are not intended for the diagnosis, prevention, treatment, cure or mitigation of any disease in humans or animals.These statements have not been evaluated by the FDA. If you have a health-related condition that requires medical attention, always consult with a licensed health care professional. Individual results may vary.

For More Information:

Earthpulse Press Incorporated
P. O. Box 201393
Anchorage, Alaska 99520 USA
Voice Mail Ordering: 1-907-249-9111

http://www.earthpulse.com

24 Hours a Day
VISA or Master Card Accepted

Dealer and Wholesale Inquiries Welcome